T0406005

Iceland's Arctic Policies and Shifting Geopolitics

Valur Ingimundarson

Iceland's Arctic Policies and Shifting Geopolitics

Embellished Promise

Valur Ingimundarson
University of Iceland
Reykjavík, Iceland

ISBN 978-3-031-40760-4 ISBN 978-3-031-40761-1 (eBook)
https://doi.org/10.1007/978-3-031-40761-1

© The Editor(s) (if applicable) and The Author(s), under exclusive license to Springer Nature Switzerland AG 2024

This work is subject to copyright. All rights are solely and exclusively licensed by the Publisher, whether the whole or part of the material is concerned, specifically the rights of translation, reprinting, reuse of illustrations, recitation, broadcasting, reproduction on microfilms or in any other physical way, and transmission or information storage and retrieval, electronic adaptation, computer software, or by similar or dissimilar methodology now known or hereafter developed.
The use of general descriptive names, registered names, trademarks, service marks, etc. in this publication does not imply, even in the absence of a specific statement, that such names are exempt from the relevant protective laws and regulations and therefore free for general use.
The publisher, the authors and the editors are safe to assume that the advice and information in this book are believed to be true and accurate at the date of publication. Neither the publisher nor the authors or the editors give a warranty, expressed or implied, with respect to the material contained herein or for any errors or omissions that may have been made. The publisher remains neutral with regard to jurisdictional claims in published maps and institutional affiliations.

This Palgrave Macmillan imprint is published by the registered company Springer Nature Switzerland AG.
The registered company address is: Gewerbestrasse 11, 6330 Cham, Switzerland

If disposing of this product, please recycle the paper.

Preface

Jean-Paul Sartre's dictum that "existence precedes essence"[1] captures the protean and contingent nature of contemporary Arctic political imaginaries. Discourses on Arctic cooperation have been juxtaposed against—and, after Russia's 2022 invasion of Ukraine, given way to—antagonist ones. Sovereignty and territorial controls, with the Arctic states seeking a regional hegemonic role, have been pitted against the interests of non-Arctic states and organized non-governmental formations. In an age of climate change—where the permeability of borders is unavoidable—the Arctic has been projected as an ecological frontier, a haven to be regulated by an international regime, or as a high-stakes resource base to be exploited for economic gain and societal needs. And within Arctic communities, a modernist Western self-determination vocabulary coexists with decolonization narratives, centering on Indigenous rights.

In this book, I explore how such tropes have influenced Iceland's Arctic political narratives from the end of the Cold War to the present. The focus is on two broad themes: First, I analyze the discursive role of the Arctic in Iceland's foreign and security policies during this period and put it within a geopolitical context, such as the end of the Cold War, the emergence of climate change on the international agenda, the global financial crisis, and the resumption of rivalry between the West and Russia. Second, I show how Icelandic elites have used the Arctic, instrumentally, for various political, economic, and cultural purposes when engaging with other states: as a post-Cold War hedging instrument after the United States ended—temporarily, as it turned out—a long-standing military presence in Iceland; as a crisis response mechanism to mitigate the effects of Iceland's financial crash during the Great Recession; as a form of identity politics in connection with Iceland's status-seeking in international forums; and as part of a nation branding exercise associated with Iceland's geostrategic location. In short, I am concerned with how Iceland has been portrayed and cast as an Arctic state, how its Arctic political engagement has manifested itself in its foreign and security policies, how other states and institutions have influenced its Arctic posture, and how its stakeholding aspirations in the region have been pursued, perceived, and received.

The project is the result of my scholarly engagement—in the past decade and a half—with Arctic geopolitics and governance as well as with Iceland's political place in the region. I am thankful to Palgrave Macmillan editor Stewart Beale for his suggestion that I turn my ideas about Iceland and Arctic politics into a monograph. I have also published articles and book chapters that reflect my thinking on the subject over time. In addition, I have had the chance to present papers on the themes of this work at international conferences and at various institutions and organizations, such as the London School of Economics, Centre for International Studies (CIS), the Royal United Services Institute (RUSI) in London, the Norwegian Institute of International Affairs (NUPI), the Norwegian Institute for Defence Studies (IDF), the Shanghai Institutes for International Studies (SIIS), the Naval War College in Newport, the National University of Singapore, the University of Greenland, the St. Petersburg University of the Humanities and Social Sciences, and the Sasakawa Peace Foundation (OPRI-SPF) in Tokyo.

I am especially grateful to the University of Iceland for providing me with research grants to enable me to work on this book. I am also indebted to research institutions that I have been affiliated with: the EDDA Research Center at the University of Iceland, which I chair; the ReNew (Reimagining Norden in an Evolving World) research hub; LSE IDEAS where I am a Research Associate; RUSI where I am a Senior Associate Fellow; and the Center for Transnational Relations, Foreign and Security Policy at the Freie Universität in Berlin (ATASP), where I spent time as a Visiting DAAD fellow. I would like to thank the publisher, Palgrave Macmillan, for a fruitful cooperation. Production editors Nirmal Kumar and Sheetal Sharma navigated me through the publishing process, with care and support; I also want to express my appreciation to Shilpa Amarpuri and Ulrike Stricker-Komba for their assistance in the editorial and publication process.

The comments by the external reviewers as well as others who have read the manuscript were helpful in improving it. Several historians and political scientists working at the National Archives of Iceland have facilitated my work, especially Benedikt Eyþórsson and Kristinn Valdimarsson. This endeavor has benefited from intellectual exchanges with colleagues, students, and friends over a period of time, including the late Christopher Coker, Rolf Tamnes, Chris Alden, Irma Erlingsdóttir, Walter Berbrick, Halla Gunnarsdóttir, Egill Þór Níelsson, Lawson Brigham, Gunnar Gunnarsson, Karl Blöndal, Yang Jian, Þorsteinn Gunnarsson, Nataly Marchenko, Magnús Jóhannesson, Jóhann Sigurjónsson, Michael Codner, Klaus Dodds, Arild Moe, Odd Jarl Borch, and Anna Karlsdóttir. Finally, I would like to thank my research assistant, Svanhildur Ástþórsdóttir, for her extensive and valuable contribution.

Reykjavík, Iceland Valur Ingimundarson

Note

1. See Jean-Paul Sartre, "Existentialism is humanism," in *Existentialism: Basic writings*, ed. Charles Guignon and Derk Pereboom (Indianapolis, IN: Hackett Press, 1946 [2001]), 290–308.

Contents

Abbreviations

AC	Arctic Council
AEC	Arctic Economic Council
AEPS	Arctic Environmental Protection Strategy
ACGF	Arctic Coast Guard Forum
AMAP	Arctic Monitoring and Assessment Programme
ASFR	Arctic Security Forces Roundtable
BEAC	Barents Euro-Arctic Council
BRI	Belt and Road Initiative
CAFF	Conservation of Arctic Flora and Fauna
CIAO	China-Iceland Arctic Science Observatory
CRI	Carbon Recycling International
CBSS	Council of the Baltic Sea States
CNOOC	China National Offshore Oil Corporation
EPPR	Emergency Prevention, Preparedness and Response
EU	European Union
EEA	European Economic Area
EEZ	Exclusive Economic Zone
EFTA	European Free Trade Association
FDI	Foreign Direct Investment
FTA	Free Trade Agreement
GIUK	Greenland-Iceland-UK [Gap]
IASC	International Arctic Science Committee
ICJ	International Court of Justice
IMF	International Monetary Fund
IMO	International Maritime Organization
IUU	Illegal, Unreported and Unregulated [Fishing]
MADIZ	Military Defense Identification Zone

MARPOL	International Convention for the Prevention of Pollution from Ships
NATO	North Atlantic Treaty Organization
ND	Northern Dimension
NORDEFCO	Nordic Defence Cooperation
NSR	Northern Sea Route
OECD	Organisation for Economic Co-operation and Development
PCA	Permanent Court of Arbitration
PCIJ	Permanent Court of International Justice
PAME	Protection of the Arctic Marine Environment
RFMO	Regional Fishing Management Organization
SAO	Senior Arctic Official
SAR	Search and Rescue
SCPAR	Standing Committee of Parliamentarians of the Arctic Region
SDI	Strategic Defense Initiative
SLOC	Sea Lines of Communication
SOLAS	International Convention for the Safety of Lives at Sea
UNCLOS	United Nations Convention on the Law of the Sea
USGS	U.S. Geological Service
WWF	World Wildlife Fund

CHAPTER 1

Introduction

The role of the Arctic in Iceland's foreign and security policies has been heavily influenced by external geopolitical factors. After World War II, Iceland was associated with Arctic militarization as part of its Cold War integration into NATO and the global U.S. base network. In the closing years of the East-West conflict, it, symbolically, became involved in efforts to break out of a Soviet-American stalemate through its hosting of the Reykjavík Summit, which was followed by Soviet leader Mikhail Gorbachev's heralding of a new era of intergovernmental cooperation in the Arctic. In the post-Cold War period, Iceland cemented its status as an Arctic state by becoming a member of several Northern subregional organizations—designed to integrate Russia into Western institutional structures—by assuming the rotating chair of the Arctic Council (AC) and by taking part in Arctic science cooperation. Yet, despite these various links to the Arctic, Iceland did not prioritize the area until two seminal events prompted it to do so: the U.S. military withdrawal from the island in 2006, relegating to history one of the Cold War's main theaters in the North Atlantic, and the collapse of the Icelandic banking system during the height of the global financial crisis in 2008.

The unilateral decision by the United States to terminate a military presence, spanning more than half a century, occurred after a protracted and bitter diplomatic dispute with Iceland, which had opposed it on the grounds that it violated contractual defense commitments.[1] To be sure,

© The Author(s), under exclusive license to Springer Nature Switzerland AG 2024
V. Ingimundarson, *Iceland's Arctic Policies and Shifting Geopolitics*, https://doi.org/10.1007/978-3-031-40761-1_1

the Americans reaffirmed their commitment to the 1951 U.S.-Icelandic Defense Agreement through "deterritorialized" defense.[2] Still, without a perceived Russian threat, they felt no need for keeping troops in Iceland permanently given their preoccupation with the Middle East, the "War on Terror"—after the attacks on September 11, 2001—and the subsequent wars in Afghanistan and Iraq. In response to the U.S. departure, the Icelandic government partly sought to make up for the loss of Iceland's strategic value by promoting itself as an Arctic state in its foreign policy. The purpose was to take advantage of increased global attention on the North, stemming from the debate over climate change and Arctic geopolitics following the Russian theatrical act, in 2007, to plant a flag on the bottom of the North Pole. Iceland's banking crash was the largest that any state had experienced relative to the size of the economy, forcing it to scramble for ways to deal with the crisis, which included attempts to forge new external relationships with countries outside the Western domain such as China.[3] What aroused Icelandic interest in the Arctic was that it was seen as a site of opportunities tied to the prospective opening of new shipping lanes and to natural resource extraction[4] in the wake of the U.S. Geological Survey's (USGS) much-quoted estimation that 13% of the world's "undiscovered" oil and 30% of its natural gas were in the region.[5] Thus, the Arctic assumed a specific function in an Icelandic post-crisis narrative articulated by political elites who were drawn to the region's future economic promise.

Together with Iceland's 2009 application for membership in the European Union (EU), the Arctic became a centerpiece of its foreign policy. Iceland made no jurisdictional claims in the Arctic Ocean—grounded in provisions of the United Nations Convention on the Law of the Sea (UNCLOS)—dealing with the delimitation of boundaries on the outer continental shelf. Its political weight was mostly confined to its membership in the Arctic Council (AC), whose emphasis was on environmental concerns and, later, on sustainable development, as well as in other less important regional intergovernmental and subregional organizations. Together with Finland and Sweden, Iceland was not counted among the five Arctic littoral states—that is, Russia, Canada, the United States (Alaska), Denmark (Greenland), and Norway—or the so-called Arctic Five, which made stakeholding claims in the Arctic Ocean.[6] Yet it belonged to an exclusive group of the eight Arctic states, which made up the Arctic Council as permanent members. While pledging loyalty to the Law of the Sea, the Arctic Eight collectively resisted outside interference in the Arctic.

Hence, they shared a profound skepticism toward proposals—espoused by non-governmental organizations and transnational bodies such as the European Parliament in the late 2000s[7]—to negotiate an Arctic Treaty like the 1959 Antarctic Treaty and to turn the region into a scientific preserve and sanctuary.[8] Such ideas would, in their views, infringe on claims in the Arctic region, which were based on the 200-mile Exclusive Economic Zone (EEZ) regime, as well as on sovereign state rights and jurisdiction beyond them.[9] True, they opened space for regional cooperation through the non-traditional proceedings of the Arctic Council, where apart from permanent member states and permanent participants from Indigenous organizations, a variety of state and non-state observers were represented. But the Arctic states were determined to maintain their control over the Arctic Council and preserve their privileged role in Arctic governance. This meant rejecting counter-discourses on an Arctic common, unsettled boundaries or fluid environments, which the Arctic Eight interpreted as having a destabilizing effect on sovereign sanctity. Any suggestions of an alternative governance model, legitimizing non-state stakeholding rights, could lead to the downgrading of state territorial functions in the Arctic.[10]

The Arctic Eight agreed on the goal to keep the Arctic a "stable" and "peaceful" region based on intergovernmental cooperation, echoing the non-military agenda of the Arctic Council. Inflated, media-driven scholarly accounts of a "Scramble for the Arctic" and predictions of Great Power rivalry following the Russian flag-planting spectacle did not shake that common position.[11] Yet many Arctic scholars disagreed with the notion that the Arctic was on the cusp of geopolitical rivalry, stressing that it did not serve the interests of the main stakeholders.[12] This is not to say that Russia's symbolic action was without consequences. It spurred intense nationalistic reaction among Canadian politicians,[13] expedited the crafting of a U.S. Presidential Directive on the Arctic[14] and paved the way for the first meeting of the Arctic Five in Ilulissat, Greenland. Still, there was no interest in dealing with the Arctic by going back to competitive nineteenth-century geopolitics symbolized by the "Scramble for Africa" during the New Imperialism period.[15] Instead, a new metanarrative emerged, which was captured in the Norwegian catchphrase "High North, low tension," portraying the region as "exceptionalist,"[16] or outside the frame of Great Power rivalries in other regions. Thus, the 2008 Ilulissat Declaration of the Arctic Five committed them to the peaceful settlement of international disputes in the Arctic.[17]

Iceland was firmly committed to the Arctic Council as the central regional body in the Arctic, and it would even have preferred a stronger institution with a treaty-mandate as originally proposed by Canada.[18] What successive Icelandic governments feared was that an institutionalization of the collaboration of the Arctic Five would lead to an alternative regional hegemonic framework.[19] This disposition reflected what the Arctic scholar Franklyn Griffiths termed "possession anxiety" in another context—the fear of losing something or, in this case, the status as a fully recognized Arctic state.[20] Thus, the Arctic Five forum was seen as undermining the Arctic Council by excluding representatives from the three other Arctic states as well as those from the organizations of the Arctic Indigenous peoples.[21] These fears were partly allayed when the authority of the Arctic Council was reasserted through the conclusion—"under its auspices"—of the two internationally binding intergovernmental agreements on Search and Rescue Agreement (SAR) in 2011 and on Marine Oil Spill Preparedness and Response in 2013, which was followed by a third one on scientific cooperation in 2017.[22] The admission of six additional countries as observers in the Arctic Council, including China, Japan, and India, in 2013, also strengthened its international legitimacy.[23]

Iceland's preoccupation with the Arctic, which reached its height in 2013, waned in the following years due to government changes, post-crisis economic developments, and geopolitical shifts. First, Iceland's surprisingly quick economic recovery, which was driven by a tourist boom and by a strong performance of core pillars of its economy—fishing and, to a lesser extent, aluminum smelting—alleviated the pressure to instrumentalize the Arctic as a source of a future economic dividend. It also made the government less keen on foreign policy experiments, such as an active engagement with China. Second, the 2014 Ukrainian crisis—and the subsequent Russian annexation of Crimea—led to the reinforcement of NATO's military posture in Europe, with direct repercussions for Iceland.

Initially, the tensions between the West and Russia over Ukraine did not have much impact on Arctic governance, such as the working of the Arctic Council or non-military cooperation of the Arctic Eight on search and rescue. After the Crimean seizure, only one Arctic Council meeting was boycotted by Western states,[24] and cooperation within the forum quickly resumed.[25] Other areas, however, were affected by the geopolitical fall-out. Apart from the suspension of military collaboration between NATO and Russia, Western economic sanctions hit Russia's oil and gas exploration in the Arctic.[26] Moreover, in a Cold War throwback, the United States and NATO renewed their military interest in the North Atlantic as well as the Arctic for the first time since the disintegration of the Soviet Union. It was justified by a need to balance Russia, with Cold War concepts like "deterrence," "collective defense," and the "GIUK gap" (Greenland–Iceland–United Kingdom gap) being resuscitated, reframed, and recycled for this purpose.[27] The rationale was to deny Russia the military capability to exploit the sea routes from the Arctic into the North Atlantic and the "GIUK gap"—the strategic choke point, representing the Russian Northern Fleet's gateway to the Atlantic Ocean.[28] It covered an area of 700 miles and was part of the U.S. Navy's anti-submarine warfare posture and its mission to monitor Soviet submarines in peacetime.

True, the maritime domain around the "GIUK gap" had changed, significantly, since the Cold War, with the Norwegian and Barents Sea being used for extensive commercial activities, scientific, exploration, and even tourism. The area had also drawn attention to states with a global reach, such as China, due to possible new shipping routes and access to energy resources.[29] But the bulk of Russia's submarine force were—and still are—located in the Kola Peninsula, which is the primary site of its second-strike nuclear capability. The Russian "bastion concept" centers on defending its nuclear forces in this wider region.[30] And to enable the Northern Fleet to have access to the Northern Sea Route (NSR) from the Atlantic Ocean to the Pacific Ocean, Russia built up military infrastructure facilities along its vast northern coast.[31]

In the Western calculus, the strategic location of Iceland made it, again, an optimal location as a base for anti-submarine operations against Russia, as was the case during the last phase of the Cold War in the 1980s. The U.S. decision—as expressed in the 2017 National Security Strategy—to single out Russia and, especially, China as "strategic competitors" on a global scale played into this military thinking. The United States not only found fault with Russia for reestablishing military bases in the Arctic; it also reaffirmed its rejection of Russian sovereignty claims over the Northern Sea Route on the grounds that the straits of the NSR were international, where the principle of "the freedom of navigation" applied.[32] The Russians, on the other hand, view them as the internal waters of Russia and subject to national rules of navigation.[33] As for China, it was more about U.S. worries about its future military capabilities in the Arctic. To the Americans, China's economic and science cooperation agreements with Arctic states, including Iceland, could be seen as a harbinger of a more extensive presence.[34] This anti-China dimension was highlighted by U.S. President Donald Trump's surprising—and unsuccessful—neocolonial offer to buy Greenland in 2019,[35] which was shelved after dumbfounded Danish and Greenlandic ministers made it clear that Greenland was not up for sale.[36]

This U.S. push for renewed geopolitical demarcation lines in the Arctic affected Iceland's security policy discourse. Iceland continued—like the other Arctic states—to refer to the Arctic as a low-tension area, where peaceful cooperation and respect for international law and sovereign interests were maintained. Nonetheless, it began to attribute increased regional tension to Russia's military buildup.[37] It also called for a NATO role in the North Atlantic as well as in the Arctic together with a deeper defense and security policy cooperation between the Nordic states. Thus, a standard Arctic cooperation narrative became mixed with a competitive one. An adversarial picture was drawn up of Russia, which turned into a hostile one following its invasion of Ukraine in 2022. Suspicions were also raised of China's geopolitical aims in the Arctic, not least based on its behavior in the South China Sea, where it has put forward highly contested sovereignty claims.[38] In short, a discourse of Arctic exceptionalism was gradually abandoned due to the return to a more polarizing geopolitical environment.

In this book, I analyze these Arctic trajectories in Iceland's foreign and security discourses and policies. The focus is on Iceland's ties with other Arctic states as well as non-Arctic states, its alliance politics with the United States and NATO, and its contribution to regional cooperation in Arctic forums and organizations, notably, the Arctic Council. Special attention will be devoted to the question of how the Arctic has been used for national identity politics, status-seeking, and branding purposes. Iceland's Arctic policies will also be examined within the context of broader geopolitical, economic, and environmental trends. These include the break-up of the Soviet Union, Arctic ice-melting and climate change, the prospects for natural resource extraction and the opening of new sea lanes and transportation routes, the impact of the global financial crisis, the involvement of China in the Arctic, the growing remilitarization of the North Atlantic and Arctic due to the deterioration in Western relations with Russia over the Crimea annexation, U.S. global rivalry with Russia and China, and Russia's invasion of Ukraine.

While employing a historical narrative approach in this study, I build on two theories in international relations—hedging and status-seeking. As a post-Cold War construct, "hedging" highlights the ambivalence of alignment strategies adopted by governments, using a mixture of cooperative and adversarial elements in their dealings with Great Powers.[39] As a middle position, it reflects selective engagement, limited resistance, and partial deference. Hedging is often contrasted with alternative Cold War concepts such as "balancing," which involves taking clear sides—for example through alliances—to address perceived state threats or "bandwagoning," which suggests deferring to, and riding the coattails of, a Great Power. In the absence of an East-West confrontation and clear-cut foreign policy choices in the 1990s and the 2000s, it was felt that the tendency of governments in small- or medium-sized states to refrain from aligning unreservedly with larger ones was not reflected in prevailing theories. New questions were raised about how to respond to rising powers, such as China, or to guard against the possibility of the abandonment of a friendly Great Power like the United States. Given China's rapid emergence as a world power in the twenty-first century, it is not surprising that the term "hedging" first appeared in political, economic, and security studies on the Asia-Pacific, where the United States had a large political and military presence. Yet, lately, it has been used to analyze developments in Europe, Eurasia, the Middle East, and other regions with a focus on "maneuvering spaces" for small states in their dealings with Great Powers.[40] For this reason, is it not surprising that both the United States and China view hedging as an opportunistic behavior. Instead, governments should send unambiguous signals of loyalty and commit to a prescribed ideological orientation—articulated, in the case of the United States, for example, by referring to the preservation of what it terms a "rule-based order," or, in the case of China, by "building a community of common destiny."[41]

In this account, hedging is used in a specific way: to account for attempts by Icelandic elites to pursue alternative alignment options in a time of rupture in the late 2000s.[42] It was not about abandoning pre-existing, state-based, or institutional relationships with Western states but about experimenting with new ones and supplementing old ones. It was not a well-thought-out or calculated strategy; it was hastily concocted and driven by desperate domestic needs and psychological pressures. After experiencing "abandonment," Iceland reached out to international organizations and Western and non-Western countries for assistance. To compensate for the U.S. military departure, it sought to reinforce its ties with

NATO through air policing agreements with member states, and it also concluded security agreements with neighboring states. After the banking collapse, it first approached Russia for an emergency loan before turning to the International Monetary Fund (IMF) for a bail-out; it, then, applied for EU membership, while, at the same time, eying China for new economic and trade opportunities.

I argue that the Arctic played an important functional role in this haphazard strategy: It symbolized Iceland's place in an area that was gaining increasing international attention and that was projected—domestically—in terms of future economic promise. In other words, Iceland, which had been wedded to the United States and Western and Northern Europe in its foreign policy during the Cold War—even if the U.S. military presence on the island was, at times, vigorously contested—engaged in hedging to deal with new geopolitical realities. It situated itself, briefly, between competing and aligned powers where it tried to navigate a broad range of risks and uncertainties. What made it easier for Iceland to follow such a strategy without interference for several years was the absence and non-engagement of its traditional patron, the United States, after it closed its military base in the country. As a short-term response to a crisis, Iceland's hedging strategy failed in many respects. But what the experience informs, in general, about hedging is its use in high-stake situations, when there is unclarity about, or non-belief in, traditional security threats and the reliability of Great Power relationships.[43] It will be shown here that hedging ceased to shape Icelandic foreign policy—without being totally abandoned—after the country's economic recovery and after the United States turned, again, its military attention on Iceland. It also reflected a broader tendency noted by scholars who have studied hedging in other contexts—namely, that the space to hedge usually shrinks with increased U.S.-China competition.[44]

The second strategy that Iceland used to promote its Arctic interests had to do with status-seeking based on social recognition: to be accepted by other governments as belonging to an Arctic club, enjoying the privileges of membership.[45] After elevating the Arctic in Icelandic foreign policy, political elites carved out "status-markers." They included Iceland's geostrategic location and its membership in the Arctic Council and other northern subregional organizations, in addition to being the site of the Arctic Circle Assembly—a high-profile annual conference project on the Arctic initiated by former Icelandic President Ólafur Ragnar Grímsson. The purpose was not only to draw attention to Iceland as part of Arctic identity politics and nation branding but also to be granted access to decision-making on Arctic issues.

Iceland's status-seeking was, initially, part of a national rehabilitation project due to the financial crisis. It reflected a need to show domestic and foreign audiences that Iceland was not only part of the Arctic but also that the region was key to its material future. Yet, following its economic revival, Iceland's status-seeking agenda became more ambitious, with claims to become recognized as an Arctic coastal state—in its 2011 Arctic policy—and as a global Arctic hub through the Arctic Circle conference. What is more, Iceland was offered as a "role-model" for sub-state Arctic communities without full decision-making powers, with a special focus on Greenland. I argue that the status-seeking strategy has had a mixed record. Iceland did not retain its Arctic coastal state demand in its revised 2021 Arctic strategy. Its involvement in negotiations initiated by the Arctic Five, which led to an agreement, in 2018, on a fishing moratorium in the Central Arctic Ocean was seen as a recognition of its stakeholding role. Still, from the perspective of the Arctic Five, another reason may have been more important: to ensure that Iceland would not try to sabotage the agreement. And while the Arctic Circle Assembly—Grímsson's personal legacy-cum-national branding project—has become a central Arctic conference venue, claims about its direct influence on governance or state policies do not hold up.

This monograph builds upon my research on the role of the Arctic in Icelandic foreign and security policy and puts it within the broader scholarly literature.[46] It is divided into three segments. First, I introduce Iceland's political place in the Arctic in the twentieth century, its Arctic identity and status-seeking projections, and its involvement in Arctic regionalization efforts after the end of the Cold War. Second, I discuss the impact of the collapse of the Icelandic banking system on Iceland's Arctic policies and its relations with other states, such as China, and analyze how the region became central in Icelandic nation branding efforts. Finally, I engage with how changes in the global geopolitical landscape, especially Russia's annexation of Crimea, the U.S.-Sino rivalry, and the Russian invasion of Ukraine, affected Iceland's geostrategic position as an Arctic state and its foreign and security policies.

When Icelandic political elites decided to put the spotlight on the Arctic in the second half of the 2000s, they offered a cultural-historical interpretation of an Icelandic past to underpin a discourse on a materialistic future. I argue that Icelandic Arctic narratives functioned, for a time, as a domestic political displacement factor. While the end of the U.S. military presence sparked an "Arctic turn" in Icelandic foreign policy, it was, primarily, in response to the financial crisis that politicians began to "seize upon" the Arctic. What further made the Arctic attractive as an instrument were political expedience and regional incompleteness. By juggling diverse political, economic, and cultural factors, the topicality of the Arctic was articulated by putting forward an ideological interpretation of the region's "future return" unencumbered by short-term policy accountability or political scrutiny. The opening of sea lanes and resources extraction in the Arctic was a long-term prospect, which went far beyond the length of government terms. Thus, the Arctic was first projected as a symbol of regional economic promise, especially after the financial crisis.[47] Within a broader policy framework, this narrative unleashed all kinds of government proposals and initiatives—under the stewardship of Foreign Minister Össur Skarphéðinsson—from 2009 to 2013. They included making Iceland a transshipment hub, establishing a search and rescue center at the former U.S. Naval base, granting a license for oil exploration in Iceland's EEZ, with the participation of Norwegian and Chinese oil companies, and concluding an Arctic framework agreement, with emphasis on science cooperation and a Free Trade Agreement with China.

Based on these trajectories, I will interpret Iceland's Arctic policies within the context of specific geopolitical shifts. I show that from 2008 to 2014, Iceland pursued an agenda, where the Arctic was treated in exceptional terms as a self-contained area mostly untainted by external conflicts or geopolitical machinations. It also fitted with Iceland's reconstruction strategy—after the financial crash—of not only prioritizing its interests in the Arctic by cultivating traditional ties with Western countries, bilaterally and through multilateral bodies, but also with expanding economic and trade relations with other countries, such as China. The 2014 Ukrainian crisis, initially, did not alter, in any significant way, Iceland's approach toward the Arctic, which continued to be couched in collaborative, non-military terms. Western tensions with Russia and the militarization of the North Atlantic were not directly tied to the Arctic, even if Iceland took full part in the sanction regime against Russia. Yet, as I make clear, coinciding with increased U.S. military presence in Iceland since 2016, the Arctic became increasingly subordinated to a backward-looking security policy focus on the North Atlantic and on the defense relationship with the United States. Thus, diachronic ties with the Cold War era were reestablished. Icelandic policymakers, like those of other Arctic states, continued to plead for a non-militarized Arctic. It did not, however, prevent them from expressing, simultaneously, hard security concerns, which were influenced by Western alliance politics in a more contentious geopolitical setting, and which became dominant after Russia's 2022 invasion of Ukraine. Hence, the role of Iceland's strategic location in regional security cooperation, notably NATO but also the Arctic Security Forces Roundtable (ASFR) and the Joint Expeditionary Force (JEF),[48] has become more important.[49] Yet, despite the breakdown in cooperation between the West and Russia in Arctic forums, the hope has not been abandoned that the regional institutional structure, which was created after the end of the Cold War, can be revived in the future.[50]

Notes

1. On the historical and contemporary dimensions of the U.S.-Icelandic military relationship, see Valur Ingimundarson, *Í eldlínu kalda stríðsins. Samskipti Íslands og Bandaríkjanna 1945–1960* [In the Crossfire: Icelandic-U.S. Relations 1945–1960] (Reykjavík: Vaka Helgafell, 1996); Ingimundarson, *Uppgjör við umheiminn: Samskipti Íslands við Bandaríkin og NATO 1960–1974* [A Reckoning with the Outside World: Iceland's Relations with the United States and NATO, 1960–1974] (Reykjavík: Vaka Helgafell, 2001), 50–54; Ingimundarson, *The Rebellious Ally: Iceland, the United States, and the Politics of Empire, 1945–2006* (Dordrecht and St. Louis: Republic of Letters, 2011); Ingimundarson, "Unarmed sovereignty versus foreign base rights: enforcing the US-Icelandic defence agreement 1951–2021," *The International History Review* 44, no. 1 (2022): 73–91; Ingimundarson, "A Fleeting or Permanent Military Presence? The Revival of US Anti-Submarine Operations from Iceland," *RUSI Newsbrief* 38, no. 7 (2018): 1–4; Gustav Pétursson, "The Defence Relationship of Iceland and the United States and the Closure of the Keflavík Base," PhD. diss. (University of Lapland, 2020); Guðni Th. Jóhannesson, "To the Edge of Nowhere?" *Naval War College Review* 57, no. 3 (2004): 115–137; Michael T. Corgan, *Iceland and Its Alliances: Security for a Small State* (New York: E. Mellen Press, 2002); Thor Whitehead, *The Ally Who Came in from the Cold: A Survey of Icelandic Foreign Policy, 1946–1956* (Reykjavík: University of Iceland Press 1998).
2. See "Samkomulag Bandaríkjanna og Íslands um aðgerðir til að styrkja varnarsamstarf ríkjanna" [Joint Understanding between the United States and Iceland on Measures to Strengthen the Bilateral Defense Cooperation], October 2006, accessed May 23, 2023, https://www.utanrikisraduneyti.is/media/Frettatilkynning/Samkomulag_um_varnarmal.pdf.

3. On the Icelandic financial crisis, see Ásgeir Jónsson and Hersir Sigurgeirsson, *The Icelandic Financial Crisis: A Study into the World's Smallest Currency Area and its Recovery from Total Banking Collapse* (London: Palgrave, 2016); Valur Ingimundarson, Philippe Urfalino, and Irma Erlingsdóttir, eds., *Iceland's Financial Crisis: The Politics of Blame, Protest, and Reconstruction* (New York and London: Routledge, 2016); Eiríkur Bergmann, *Iceland and the International Financial Crisis: Boom, Bust and Recovery* (New York: Palgrave Macmillan, 2014); Robert Z. Aliber and Gylfi Zoega, eds., *Preludes to the Icelandic Financial Crisis* (New York: Palgrave Macmillan, 2011); Guðrún Johnsen, *Bringing Down the Banking System: Lessons from Iceland* (New York: Palgrave Macmillan, 2014); Ásgeir Jónsson, *Why Iceland? How One of the World's Smallest Countries Became the Meltdown's Biggest Casualty* (New York: McGraw-Hill, 2009); E. Paul Durrenberger and Gisli Palsson, eds., *Gambling Debt: Iceland's Rise and Fall in the Global Economy* (Boulder: University of Colorado Press, 2014); Guðni Th. Jóhannesson, *Hrunið* [The Crash] (Reykjavík: JPV, 2009); Roger Boyes, *Meltdown Iceland: How the Global Financial Crisis Bankrupted an Entire Country* (London: Bloomsbury, 2009).
4. About early Northern discourses in the first half of the 2000s, see Lassi Heininen and Heather N. Nicol, "The Importance of Northern Dimension Foreign Policies in the Geopolitics of the Circumpolar North," *Geopolitics* 12, no. 1 (2007): 133–165.
5. See U.S. Geological Survey, *Circum-Arctic Resource Appraisal: Estimates of Undiscovered Oil and Gas North of the Arctic Circle* (Washington, D.C.: U.S. government, 2008), accessed November 15, 2022, https://pubs.usgs.gov/fs/2008/3049/.
6. See Jon Rahbek-Clemmensen and Gry Thomasen, "How has Arctic coastal state cooperation affected the Arctic Council?" *Marine Policy* 122 (2020): 1–7, https://doi.org/10.1016/j.marpol.2020.104239.
7. See Jon Rahbek-Clemmensen and Gry Thomasen, "Learning from the Ilulissat Initiative: State Power, Institutional Legitimacy, and Governance in the Arctic Ocean 2007–1" (Centre for Military Studies, University of Copenhagen, 2018), accessed February 9, 2024, https://cms.polsci.ku.dk/publikationer/learning-from-the-ilulissat-iniative/download/CMS_Rapport_2018__1_-_Learning_from_the_Ilulissat_initiative.pdf; see also Oran R. Young, "Whither the Arctic? Conflict or cooperation in the circumpolar north," *Polar Record* 45, no. 1 (2009): 73–82; "The Ilulissat Declaration" issued by Arctic states at the Arctic Ocean Conference in Ilulissat, Greenland, May 27–29, 2008, accessed February 25, 2024, https://referenceworks.brillonline.com/entries/international-law-and-world-order/ve5-the-ilulissat-declaration-on-the-arctic-ocean-SIM_032765.

8. European Parliament Resolution on Arctic governance, October 9, 2008, accessed February 7, 2024, https://www.europarl.europa.eu/doceo/document/TA-6-2008-0474_EN.html.
9. See, for example, Bjarni Már Magnússon, *The Continental Shelf Beyond 200 Nautical Miles: Delineation, Delimitation and Dispute Settlement* (Leiden and Boston: Brill Martinus Nijhoff, 2015).
10. See Timo Koivurova, Pirjo Kleemola-Juntunen and Stefan Kirchner, "Arctic Regional Agreements and Arrangements," in *Research Handbook on Polar Law*, ed. Karen N. Scott and David L. VanderZwaag (Cheltenham: Edward Elgar Publishing, 2020), 64; Oran R. Young, "Whither the Arctic? Conflict or cooperation in the circumpolar north."
11. See Scott G. Borgerson, "Arctic Meltdown: The Economic and Security Implications of Global Warming," *Foreign Affairs* 87, no. 2 (March/April 2008): 63–77; Borgerson, "The Great Game Moves North," *Foreign Affairs*, March 25, 2009, accessed September 30, 2022, https://www.foreignaffairs.com/articles/commons/2009-03-25/great-game-moves-north; Richard Sale and Eugene Potapov, *The Scramble for the Arctic* (London: Frances Lincoln, 2010); Barry Zellen, *Arctic Doom, Arctic Boom* (London: Praeger, 2009); Alun Anderson, *After the Ice: Life, Death, and Geopolitics in the New Arctic* (New York: Smithsonian Books, 2009); Michael Byers, *Who Owns the Arctic? Understanding Sovereignty Disputes in the North* (Vancouver: Douglas and Mclntyre, 2009); Charles Emmerson, *The Future History of the Arctic* (London: The Bodley Head, 2010); David Fairhall, *Cold Front: Conflict Ahead in Arctic Waters* (London and New York: I. B. Tauris, 2010); Roger Howard, *The Arctic Gold Rush: The New Race for Tomorrow's Natural Resources* (London and New York: Continuum, 2009).
12. See, for example, Oran R. Young, "The future of the Arctic: cauldron of conflict or zone of peace?" *International Affairs* 87 no. 1 (2011): 185–193; Klaus Dodds, "The Ilulissat Declaration (2008): The Arctic States, 'Law of the Sea,' and Arctic Ocean," *SAIS Review of International Affairs* 33, no. 2 (2013): 45–55; Timo Kouvurova, "How to Improve Arctic International Governance," *U.C. Irvine Law Review* 8 (2016): 83–98; Hans Corell, "The Arctic: An Opportunity to Cooperate and to Demonstrate Statesmanship," *Vanderbilt Journal of Transnational Law* 42 (2009): 1065–1079; Frédéric Lasserre, Jérôme Le Roy, and Richard Garon, "Is There an Arms Race in the Arctic," *Centre of Military and Strategic Studies*, 14, 3–4 (2012): 1–56.
13. See, for example, "Canada rejects Arctic flag-planting as 'just a show by Russia'," *The Sidney Morning Herald*, August 3, 2007, accessed January 10, 2024, https://www.smh.com.au/world/canada-rejects-arctic-flagplanting-as-just-a-show-by-russia-20070803-r86.html.

14. U.S. Government, *Directive on Arctic Region Policy*, January 9, 2009, accessed February 9, 2024, PPP-2008-book2-doc-pg1545 (1).pdf.
15. See, for example, Muriel Evelyn Chamberlain, *The Scramble for Africa* (London, New York: Routledge, 2013); Mieke van der Linden, *The Acquisition of Africa (1870–1914): The Nature of International Law* (Boston: Brill, 2016); Ronald Robinson, John Gallagher, and Alice Denny, *Africa and the Victorians: The official mind of imperialism* (London: Macmillan, 1961).
16. See Heather Exner-Pirot and Robert Murray, *Regional Order in the Arctic: Negotiated Exceptionalism,* The Arctic Institute, October 24, 2017, accessed November 15, 2022, 47–63, https://www.thearcticinstitute.org/regional-order-arctic-negotiated-exceptionalism/; see also Juha Käpylä and Harri Mikkola, "On Arctic Exceptionalism," The Finnish Institute of International Affairs, Working Paper 85 (April 2015), 1–22, accessed July 3, 2023, https://www.fiia.fi/wp-content/uploads/2017/01/wp85.pdf; Juha Käpylä and Harri Mikkola, "Contemporary Arctic Meets World Politics: Rethinking Arctic Exceptionalism in the Age of Uncertainty," in *The Global Arctic Handbook*, ed. Matthias Finger and Lassi Heininen, (Cham: Springer, 2019), 153–69; Daria Shvets and Kamrul Hossain, "The Future of Arctic Governance: Broken Hopes for Arctic Exceptionalism," *Current Developments in Arctic Law* 10 (2022): 49–63; Pavel Devyatki, "Arctic exceptionalism: a narrative of cooperation and conflict from Gorbachev to Medvedev and Putin."
17. Lassi Heininen, Karen Everett, Barbora Padrtova, and Anni Reissell, *Arctic policies and strategies—analysis, synthesis, and trends* (Laxenburg, Austria: International Institute for Applied Systems Analysis, 2020); P. Whitney Lackenbauer and Suzanne Lalonde, eds., *Breaking the Ice Curtain? Russia, Canada, and Arctic Security in a Changing Circumpolar World* (Calgary: Canadian Global Affairs Institute, 2019); Maria L. Lagutina, *Russia's Arctic Policy in the Twenty-first Century: National and International Dimensions* (Lanham: Lexington Books, 2019).
18. See the comments by Foreign Minister, Halldór Ásgrímsson, in *Tíminn*, November 26, 1994.
19. See *Þingsályktun um stefnu Íslands í málefnum norðurslóða* [Parliamentary Resolution on Iceland's Policy on the Arctic], *Þingtíðindi*, May 19, 2011, accessed February 11, 2023, https://www.althingi.is/altext/139/s/1148.html; interviews with a former Icelandic minister, November 9, 13, 28, 2011.
20. Franklyn Griffiths was referring to expressions of Canadian Arctic nationalism, especially the rhetoric of Canadian Prime Minister Steven Harper about the need for active presence in the Northwest Passage—the maritime route between the Arctic and Pacific Oceans—to exercise and main-

tain sovereignty over it. Iceland's approach was similar, even if it was not presented in such overly nationalistic terms. See Griffiths, "Towards a Canadian Arctic Strategy," *Zeitschrift für öffentliches Recht und Völkerrecht* (*ZaöRV*) 69 (2009): 579–624.

21. Össur Skarphéðinsson, *Þingræða* [Parliamentary speech], September 3, 2010, accessed February 9, 2024, *Þingtíðindi*, https://www.althingi.is/altext/raeda/138/rad20100903T111357.html.
22. Arctic Council, "Agreement on Cooperation on Aeronautical and Maritime Search and Rescue in the Arctic," May 12, 2011, accessed January 12, 2024, https://oaarchive.arctic-council.org/items/9c343a3f-cc4b-4e75-bfd3-4b318137f8a2; Arctic Council, "Agreement on Marine Oil Pollution Preparedness and Response in the Arctic," May 15, 2013, accessed February 1, 2024, https://oaarchive.arctic-council.org/items/ee4c9907-7270-41f6-b681-f797fc81659f; Arctic Council, "Agreement on Enhancing International Arctic Scientific Cooperation," May 11, 2017, accessed February 1, 2024, https://oaarchive.arctic-council.org/items/9d1ecc0c-e82a-43b5-9a2f-28225bf183b9; see also Svein Vigeland Rottem, "A Note on the Arctic Council Agreement," *Ocean Development & International Law* 46, no. 1 (2015): 50–59; Paul Arthur Berkmann, Alexander N. Vylegzhanin, and Oran R. Young, "Application and interpretation of the Agreement on Enhancing International Arctic Scientific Cooperation," *Moscow Journal of International Law* 49, no. 3 (2017): 6–17.
23. See Valur Ingimundarson, "Managing a contested region: the Arctic Council and the politics of Arctic governance," *Polar Journal* 4, no. 1 (2014): 183–198; see also Philip E. Steinberg and Klaus Dodds. "The Arctic Council after Kiruna," *Polar Record* 51, no. 1 (2015): 108–110.
24. Barry Scott Zellen, "As War in Ukraine Upends a Quarter Century of Enduring Arctic Cooperation, the World Needs the Whole Arctic Council Now More Than Ever," *Northern Review* 54 (2023): 137–160.
25. Interview with a former high-ranking Arctic official, February 19, 2024.
26. Interview with a high-ranking U.S. State Department official, June 30, 2014.
27. See Valur Ingimundarson, "A Fleeting or Permanent Military Presence?"
28. See, for example, Julienne Smith and Jerry Hendrix, *Forgotten Waters: Minding the GIUK Gap* (Washington, D.C.: Center for a New American Security, May 2017), accessed November 30, 2022, https://www.cnas.org/publications/reports/forgotten-waters.
29. Magnus Nordenmann, "Back to the Gap," *The RUSI Journal* 162, no. 1 (2017): 24.
30. See Rebecca Pincus, "Towards a new Arctic: Changing Strategic Geography in the GIUK Gap," *The RUSI Journal* 165, no. 3 (2020): 50–58.
31. Jonas Kjellén, "c," *Arctic Review on Law and Politics* 13 (2022): 34–52.

32. *See* U.S. Department of State, *Limits in the Seas, No. 112, United States Responses to Excessive National Maritime Claims* (Bureau of Oceans and International Environmental and Scientific Affairs, 1992), 71–73; see also Christopher R. Rossi, "The Northern Sea Route and the Seaward Extension of Uti Possidetis (Juris)," *Nordic Journal of International Law* 83, no. 4 (2014): 476–508.
33. See, for example, A. Schneider, "Northern Sea Route: A Strategic Arctic Project of the Russian Federation," *Problems of Economic Transition* 60, no. 1–3 (2018): 195–202; Jan Jakub Solski, "The Northern Sea Route in the 2020s: Development and Implementation of Relevant Law," *Arctic Review of Law and Politics* 11 (2000): 383–410; Leonid Tymchenko, "The Northern Sea Route: Russian Management and Jurisdiction over Navigation in Arctic Seas," in *The Law of the Sea and Polar Maritime Delimitation and Jurisdiction*, ed. Alex G. Oude Elferink and Donald R. Rothwell (Leiden: Brill, 2001), 277–81.
34. Address by U.S. Secretary of State Mike Pompeo at the Arctic Council Ministerial Meeting in Rovaniemi, Finland, May 6, 2018, accessed January 25, 2023, https://www.c-span.org/video/?460478-1/secretary-state-pompeo-warns-russia-china-arctic-policy-address-finland.
35. See, for example, "'Greenland Is Not for Sale': Trump's Talk of a Purchase Draws Derision," *New York Times*, August 16, 2019, accessed November 30, 2022, https://www.nytimes.com/2019/08/16/world/europe/trump-greenland.html.
36. See, for example, "No Joke: Trump Really Wants to Buy Greenland," *National Public Radio* (*NPR*), August 19, 2019, accessed May 30, 2023, https://www.npr.org/2019/08/19/752274659/no-joke-trump-really-does-want-to-buy-greenland.
37. See *Tillögur nefndar um endurskoðun á stefnu Íslands í málefnum* norðurslóða [Proposals of a [Parliamentary] Committee on the Revision of Iceland's Arctic Policy], March 8, 2021, accessed June 25, 2023, https://www.stjornarradid.is/library/04-Raduneytin/Utanrikisraduneytid/PDF-skjol/Skilabréf%20og%20tillögur%20nefndar%20um%20endurskoðun%20norðurslóðastefnu.pdf.
38. *Tillögur nefndar um endurskoðun á stefnu Íslands í málefnum norðurslóða*, 4.
39. On hedging, see John D. Ciorciari, "The variable effectiveness of hedging strategies," *International Relations of the Asia-Pacific* 19, no. 3 (2019): 523–555; Ciorciari and Jürgen Haacke, "Hedging in international relations: an introduction," *International Relations of the Asia-Pacific* 19, no. 3 (2019): 367–374; Haacke, "The concept of hedging and its application to Southeast Asia: a critique and a proposal for a modified conceptual and methodological framework," *International Relations of the Asia-Pacific* 19, no. 3 (2019): 375–417; Peter Crompton, "Hedging in academic writ-

ing: some theoretical problems," *English for Specific Purposes* 16, no. 4 (1997): 271–287; Alistair Iain Johnston and Robert S. Ross, eds. *Engaging China: The Management of an Emerging Power* (London and New York: Routledge, 1999); Alexander Korolev, "Shrinking room for hedging: system-unit dynamics and behavior of smaller powers," *International Relations of the Asia-Pacific* 19. no. 3 (2019): 419–452; Cheng-Chwee Kuik, "How do weaker states hedge? Unpacking ASEAN States' alignment behavior towards China," *Journal of Contemporary China* 25, no. 100 (2016): 500–514; Cheng-Chwee Kuik, "Getting hedging right: a small state perspective," *China International Strategy Review* no. 2 (2021): 300–310; Joshua Kurlantzik, *Charm Offensive: How China's Soft Power is Transforming the World* (New Haven: Yale University Press, 2007); Randall L. Schweller, "Bandwagoning for Profit: Bringing the Revisionist State Back In," *International Security* 19, no. 1 (1994): 72–107; Brock Tessman, "System Structure and State Strategy: Adding Hedging to the Menu," *Security Studies* 21, no. 2 (2012): 192–231; Øystein Tunsjø, *Security and Profit in China's Energy Policy: Hedging against Risk* (New York: Columbia University Press, 2013); Øystein Tunsjø, "U.S.-China Relations: From Unipolar Hedging to Bipolar Balancing," in *Strategic Adjustment and the Rise of China: Power and Politics in East Asia*, ed. Robert. S. Ross and Tunsjø (Ithaca/London: Cornell University Press, 2017), 41–68; Rosemary Foot, "Chinese strategies in a US-hegemonic global order: accommodating and hedging," *International Affairs* 82, no. 1 (2006): 77–94; Alexander Korolev, "Systemic balancing and regional hedging: China-Russia relations," *Chinese Journal of International Politics* 9, no. 4 (2016): 375–397; Ann Marie Murphy, "Great Power Rivalries, Domestic Politics and Southeast Asian Foreign Policy: Exploring the Linkages," *Asian Security* 13, no. 3 (2017): 165–182.

40. See Ciorciari and Haacke, "Hedging in international relations: an introduction," 367–368.
41. Kuik, "Getting hedging right: a small state perspective," 301.
42. Kuik, "Getting hedging right: a small state perspective," 302.
43. Kuik, "Getting hedging right: a small state perspective," 306.
44. See Korolev, "Shrinking room for hedging: system-unit dynamics and behavior of smaller powers," 419–452.
45. On status-seeking, see Marina G. Duque, "Recognizing International Status: A Relational Approach," *International Studies Quarterly* 62, no. 3 (2018): 577–592; David Lake, *Hierarchy in International Relations* (Ithaca: Cornell University Press, 2009); T.V. Paul, Deborah Welch Larson, and William C. Wohlforth, eds., *Status in World Politics* (Cambridge: Cambridge University Press, 2014); Benjamin De Carvalho and Iver Neumann, "Small states and status," in *Small state status seeking*.

Norway's Quest for International Standing, ed. Neumann and De Carvalho (London: Routledge, 2015), 56–72; Rasmus Brun Pedersen, "Bandwagon for Status: Changing Patterns in Nordic States Status-seeking Strategies," *International Peacekeeping* 25, no. 2 (2018): 217–241; Jonathan Renshon, *Fighting for Status: Hierarchy and Conflict in World Politics* (Princeton: Princeton University Press, 2017); William C. Wohlforth, Benjamin De Carvalho, Halvard Leira, and Iver Neumann, "Moral authority and status in International Relations: Good states and the social dimension of status seeking," *Review of International Studies* 44, no. 3 (2017): 526–546.

46. On Iceland's Arctic policies, see Valur Ingimundarson, "Iceland as an Arctic State," in *The Palgrave Handbook of Arctic Policy and Politics*, ed. Ken S. Coates and Carin Holroyd (London: Palgrave Macmillan, 2019), 251–265; Ingimundarson, "Framing the national interest: the political uses of the Arctic in Icelandic foreign and domestic policies," *Polar Journal* 5, no. 1 (2015): 81–100; Ingimundarson and Klaus Dodds, "Territorial nationalism and Arctic geopolitics: Iceland as an Arctic coastal state," *Polar Journal* 2, no. 1 (2012): 21–37; Ingimundarson, "Territorial Discourses and Identity Politics: Iceland's Role in the Arctic," in *Arctic Security in an Age of Climate Change*, ed. James Kraska (Cambridge: Cambridge University Press, 2011), 174–189; Ingimundarson, "Iceland's Post-American Security Policy, Russian Geopolitics and the Arctic Question," *RUSI Journal* 154, no. 4 (2009): 74–81; Alyson J.K. Bailes and Lassi Heininen, *Strategy Papers on the Arctic or High North: a comparative study and analysis* (Reykjavík: Centre for Small States, 2012); Alyson J.K. Bailes, Margrét Cela, Katla Kjartansdóttir, and Kristinn Schram, "Iceland: small but central," in *Perceptions and Strategies of Arcticness in Sub-Arctic Europe*, ed. Toms Rostoks and Andris Sprūds (Riga: Latvian Institute of International Affairs, 2014), 75–97; Margrét Cela, "Iceland: A Small Arctic State Facing Big Arctic Changes," *The Yearbook of Polar Law Online* 5, no. 1 (2013): 75–92; Margrét Cela and Pia Hansson, "A challenging chairmanship in turbulent times," *Polar Journal* 11, no. 1 (2021): 43–56; on Icelandic military security in the Arctic, see Valur Ingimundarson, "Unarmed sovereignty versus foreign base rights: enforcing the US-Icelandic defense agreement 1951–2021"; Ingimundarson, "A Fleeting or Permanent Military Presence?" Ingimundarson, "The Revival of US Anti-Submarine Operations from Iceland"; Gustav Pétursson, "The Defence Relationship of Iceland and the United States and the Closure of the Keflavík Base"; Ingimundarson, *The Rebellious Ally: Iceland, the United States, and the Politics of Empire*; Page Wilson and Auður H. Egilsdóttir, "Small State, Big Impact? Iceland's First National Security Policy," in *Routledge Handbook of Arctic Security*, ed. Gunhild Hoogensen Gjørv, Marc Lantinga, and Horatio Sam-Aggrey (Abingdon and New York:

Routledge, 2020), 188–197; Pia Hansson and Guðbjörg Ríkey Th. Hauksdóttir, "Iceland and Arctic Security: US Dependency and the Search for an Arctic Identity," in *On Thin Ice? Perspectives on Arctic Security*, ed. Duncan Depledge and P. Whitney Lackenbauer (Ontario: North American and Arctic Defense and Security Network (NAADSN), 2021), 163–171; on Icelandic soft security policies, see Ingimundarson and Halla Gunnarsdóttir. "The Icelandic Sea Areas and Activity Level up to 2025," in *Maritime activity in the High North—current and estimated level up to 2025*, ed. Odd Jarl Borch et al. (Bodø: Nord University, 2016), 74–87; Borsch, Natalia Andreassen, Svetlana Kuznetsova, Valur Ingimundarson, and Uffe Jakobsen, "Navigation safety and risk assessment in the High North," in *Marine Navigation and Safety of Sea Transportation Proceedings of the International Conference on Marine Navigation and Safety of Sea Transportation*, ed. Adam Weintritt (London: Francis & Taylor Group, 2017), 275–281.

Part I: The Background: Iceland's Role in the Arctic

47. See Valur Ingimundarson, "Framing the national interest: the political uses of the Arctic in Icelandic foreign and domestic policies."
48. JEF, a British initiative launched in 2014, is a multinational defense framework, covering the Arctic, North Atlantic, and the Baltic Sea Region, with the participation of the United Kingdom, Denmark, Estonia, Finland, Iceland, Latvia, Lithuania, the Netherlands, Norway, and Sweden.
49. Interview with a high-level Icelandic defense official, February 23, 2024.
50. Interview with a high-ranking Icelandic official, April 12, 2023; interview with a high-ranking Arctic official February 6, 2024.

CHAPTER 2

The Background: Iceland's Role in the Arctic

Historically, Iceland's interests in the Arctic have been overshadowed by those of the North Atlantic—the main source of its abundant marine resources. A national identity projection, espoused by political and cultural elites, mixed nationalist and racial ideas with a portrayal of Iceland as a "developed" European country. A "southern" look toward the European continent always took precedence over territorial aspirations in the Arctic. In this regard, there was no gap between elite and popular perceptions. It had taken Icelanders centuries to escape external stigmas of being primitive and poor, belonging to the northern edges of the inhabitable; it was not until the middle of the nineteenth century and the early twentieth century that a more positive image emerged of an "educated Iceland" and the notion of the "Hellas of the North," with references to the medieval saga literature.[1] That also meant that Icelanders did not, in any way, want to be associated with Indigeneity or with the Inuit of Greenland; on the contrary, if they showed interest in the North, it was more about colonial mimicry.[2] Thus, one academic, with some political support, argued in the first half of the twentieth century that Iceland should not recognize Denmark's colonial control over Greenland and make, instead, a territorial claim to it—a proposition that was based on spurious historical grounds.[3] What gave such a view, temporarily, more weight was Norway's legal challenge to Denmark's sovereign rights over Eastern Greenland, which the Permanent Court of International Justice (PCIJ) eventually rejected in

© The Author(s), under exclusive license to Springer Nature Switzerland AG 2024

V. Ingimundarson, *Iceland's Arctic Policies and Shifting Geopolitics*,
https://doi.org/10.1007/978-3-031-40761-1_2

1933. The Greenland case, to be sure, was never pursued seriously by Icelandic governments in the 1930s and 1940s.[4] Nonetheless, it reflected a persistent chauvinistic streak within Icelandic nationalism.

When Iceland became fully sovereign in 1918—after having been a Danish dependency for centuries—it was free to pursue its own neutralist foreign and defense policy despite maintaining a contractual relationship through the person of the Danish King.[5] After Iceland severed its remaining ties with Denmark and became a republic in 1944, one could sometimes detect a tension between a preponderant commitment to a "developed" Western nationalist trajectory and an anti-colonial narrative in its foreign policy. From the 1950s to the 1970s, these two tropes of nationalism surfaced, concurrently, in the so-called Cod Wars against the British over the unilateral extension of Iceland's fishery limits, which led to political tensions between the two NATO Allies and to skirmishes between the British Navy and the Icelandic Coast Guard. On the one hand, Iceland was influenced by standard Western nationalism based on sovereign concerns about securing its borders, especially the fishery grounds. Its policy was anti-traditional, however, because it was embedded in a modernity discourse, which was by definition ambivalent in its origin. Again, it represented an attempt to do away with notions of national "backwardness" or Indigeneity. On the other hand, Icelanders sometimes mobilized around an anti-Western narrative—based on "Third World" or Global South nationalism—against foreign rule and colonialism. This line of argument proved powerful and effective during the "Cod Wars" against a Great Power adversary, even if its credibility was dented by the fact that Iceland was an affluent country that identified with the Global North. What made it politically viable—in discursive terms—was that the Icelandic economy was existentially dependent on fishing during this period and that Iceland was fighting Britain with its imperialist and colonial past.[6]

Great Power interest in Iceland stemming from its geostrategic position in the North developed in the first half of the twentieth century. It was highlighted by Britain's occupation of the island in 1940, which was meant to prevent Nazi Germany from establishing a foothold in the middle of the North Atlantic. A year later, in July 1941, the United States, then still a non-belligerent, assumed a preponderant role in Iceland after concluding a Defense Agreement with the Icelandic government, while the British retained thousands of troops there. Because of its geostrategic location, Iceland contributed significantly to Allied military cooperation during World War II.[7] Thus, the Arctic convoys, made up of ships from the U.S., Canadian, and British Navies, sailed from Iceland to the northern ports of the Soviet Union in Arkhangelsk and Murmansk—from 1941 to 1945—providing material assistance to the Soviets prior to and after the opening of a second front in 1944.[8]

During the Cold War, Iceland also became a highly valued northern outpost for the United States and NATO through the operation of the Keflavík military base in the southwest of the country. From the late 1940s until the mid-1950s, the United States saw Iceland having a greater offensive potential than any other area except for the United Kingdom; for U.S. defense, only Greenland was considered more significant. While the introduction of intercontinental ballistic missiles and sea-launched intermediate ballistic missiles reduced the strategic importance of Iceland, it continued to be an important transatlantic link in air and sea communications between the United States and Europe.[9] In the 1950s and 1960s, a series of radar stations, constituting the so-called the Distant Early Warning Line, or DEW Line, was established across the Arctic, from Alaska through Canada over Greenland to Iceland to detect Soviet bombers coming over the North Pole. Iceland became a part of the Western "GIUK gap" strategy, with its aim of preventing Soviet submarines entering the Atlantic from hitting targets in the United States.[10] It was also a link in the U.S. Sound Surveillance System (SOSUS), a chain of underwater listening posts in the North for tracking Soviet submarines.

These anti-submarine operations precipitated a Soviet naval pivot to the Arctic in the 1970s and 1980s. The ice-free naval bases on the Kola Peninsula gave Russian surface vessels and submarines access to the North Atlantic.[11] In the first half of the 1980s, Iceland played an important role in the "forward-defense strategy" pursued by the Reagan Administration, where U.S. anti-submarine warfare forces operated and where U.S. aircraft monitored and intercepted Russian bomber planes in a specifically designated "military defense identification zone" (MADIZ) well beyond the island's 12-mile airspace.[12]

In the late 1980s, Iceland took on a new role, with improved East-West relations. In contrast to its function as a base for U.S. military activities, it became associated with disarmament in the waning years of the Cold War. By hosting the 1986 Reagan-Gorbachev Reykjavík Summit—where revolutionary ideas, such as the abolition of nuclear weapons were discussed[13]—Iceland suddenly became a symbolic venue for peace befitting a country without a military of its own. It is true that the failure to reach an agreement at the meeting due to U.S. unwillingness to give up the Strategic Defense Initiative (SDI) or "Star Wars" program led to deep disappointment. Yet "Reykjavik" was retrospectively considered, in Reagan's words, a "major turning point in the quest for a safe and secure world," paving the way for major disarmament treaties between the United States and the Soviet Union in the last years of the Cold War.[14]

2.1 The Case for an Arctic Cooperation Regime After the End of the Cold War

The Reykjavík Summit was still on Mikhail Gorbachev's mind when he delivered his noted "Murmansk Initiative" speech in October 1987. In it, he stated that Reykjavík had become a "symbol of hope that nuclear weapons are not eternal evil and that mankind is not doomed to live under the sword of Damocles."[15] While Gorbachev's intervention turned into a staple of an Arctic cooperation discourse in the 1990s and 2000s, it was rooted in a period of transition. Thus, he criticized the Pentagon's Arctic strategy for militarizing the region and reiterated a long-standing Soviet idea of a nuclear-free zone in Europe. What made his speech different, however, was that he was prepared to transcend stereotypical geopolitical thinking by offering an alternative vision of the Arctic as a "zone of peace," where intergovernmental political collaboration as well as transnational science programs should be promoted. This included support for the environmental protection of the North and for the opening of the Northern Sea Route (NSR)—controlled by the Soviet Union—for foreign ships.[16]

Gorbachev's proposition provided the backdrop of a new agenda in the Arctic by mixing together disparate elements such as peaceful resolution of disputes, resource exploitation, environmental protection, science cooperation, and the economic potential of new shipping lanes. The subsequent mythologization of the speech has largely been detached from its specific historical context—the last phase of the Cold War.[17] Thus, it is often projected, in timeless and idealized terms, as having a life of its own and underpinning an Arctic exceptionalist discourse long after Gorbachev left the political scene in 1991. There was no place in this narrative for Russian President, Boris Yeltsin, who was in power until the end of the 1990s and who did not pay much attention to the Arctic, its economic potential, or social problems.[18] After the collapse of the Soviet Union, the Arctic was seen more as a financial burden than as a site of wealth or a geopolitical asset, even if Yeltsin oversaw Russia's integration into subregional organizations. During this period, Russia ordered the abandonment of industrial sites and other types of infrastructure along the shores of the Arctic Ocean and in its North.[19] The evocation of Gorbachev's initiative did not either square with the more geopolitically assertive, economically centered, and nationalistic approach of Yeltsin's successor, Vladimir Putin, toward the Arctic in the late 2000s or with the revamping of former Russian military bases in the region.[20] True, the speech can be seen as a harbinger of Western-Russian Arctic cooperation initiatives in the 1990s. Despite Putin's military intervention in Georgia in 2008—which led to Russia's recognition of the secessionist Georgian regions of Abkhazia and South Ossetia as independent states—and the annexation of Crimea six years later, he continued to subscribe to a collaborative narrative on the Arctic. The Ukrainian crisis, however, raised questions about the continued viability of Arctic exceptionalism. And following Russia's 2022 invasion of Ukraine, the concept lost much of its resonance after the Western Arctic states suspended most cooperation with Russia in the region.

The initial Western reaction to Gorbachev's speech in 1987 had been mixed and tinged with caution at a time when the outcome of a Cold War thaw was still uncertain. The United States, which bore the brunt of the Soviet leader's criticism, saw nothing new in its message or anything that it could support on security issues. It welcomed, however, the call for increased collaboration in non-security areas such as scientific research in the Arctic.[21] While being non-committal, the Canadians refrained from criticizing the speech, characterizing it, in a minimalist fashion, as being interesting. The reaction of the Norwegians and the Finns was more openly positive.[22] Gorbachev had, after all, specifically welcomed Finnish President Mauno Koivisto's proposal for restricting naval activity in Northern sea areas—a proposal that President Reagan later rejected on the grounds that it was inappropriate to regulate U.S. naval and air activities in the context of what he termed a "regional security regime" that was "global in nature."[23] This attitude reflected U.S. insistence on its longtime foreign policy principle of the "freedom of the seas" or its interpretation of the maritime right to travel in international waters in peace and war. The Soviet leader had also mentioned Norway and Finland as potential partners in developing oil and gas deposits in the Arctic. What the Finns and Norwegian, however, were most interested in was the call for greater cooperative environmental action between the Soviet Union and neighboring countries. It was followed by the establishment of a Norwegian-Soviet Commission on Environmental Protection in 1988, which set the stage for the institutionalized partnership initiatives with Russia in the post-Cold War years,[24] with an emphasis on cleaning up Soviet-era nuclear waste, not least from the Northern Fleet.[25]

The centrist Icelandic Foreign Minister Steingrímur Hermannsson, who earlier as Prime Minister had been the host to Reagan and Gorbachev at the Reykjavík Summit, singled out the Soviet leader's ideas—in the Murmansk speech—about direct talks between North Atlantic Treaty Organization (NATO) and the Warsaw Pact on Arctic issues, about environmental protection, and about the opening of the Northern Sea Route. Still smitten by the global publicity generated by the Soviet-American encounter, Hermannsson also sought to link Gorbachev's speech to his own nation branding idea of making Iceland a hub for top-level deliberations on global problems.[26] The emphasis on the positive aspects of Gorbachev's initiative was challenged by some Icelandic right-wing critics who argued that it represented an attempt to divide the Nordic NATO countries by neutralizing the Arctic and to undermine the Alliance's defense posture in the region.[27] Generally, the speech was, however, welcomed in Iceland as signaling a commitment to a relaxation of East-West tensions and, later, economic opportunities through the opening of the Northern Sea Route.[28]

2.2 A True Arctic State? Iceland's Inclusion in Regional Forums in the North

In the immediate post-Cold War period, Icelandic governments paid scant attention to Arctic affairs. Apart from being conscious of Iceland's position as an external "strategic object" because of its location, politicians did not see the region as a privileged area. Yet, when direct economic interests were at stake, they were prepared to look further North. In 1981, Iceland and Norway agreed, for example, on the boundaries of the island of Jan Mayen, which is located about 800 kilometers from Iceland,[29] and 25 years later, in 2006, the two sides—together with the Faroe Islands and Denmark—also settled the northern continental shelf boundaries beyond 200 miles.[30] On the other hand, Iceland did not bother to become a party to the 1920 Spitsbergen Treaty until 1994 and, then, only as part of its efforts to bolster its legal position in a contentious dispute with Norway over fishing rights in the Barents Sea,[31] which was partly based on its non-recognition of the Norwegian claim to a 200-mile "Fishery Protection Zone" around Svalbard.[32]

The limited interest Icelandic elites showed in the Arctic reflected broader external trends. After the disintegration of the Soviet Union, Great Power strategic and military interest in the Arctic largely disappeared, and it was seen as a peripheral issue in European politics.[33] Other priorities, such as the European integration and enlargement processes, NATO's expansion, and the wars in the former Yugoslavia became the center of attention. From the 1990s until the mid-2000s, Iceland was preoccupied with its ultimately unsuccessful effort, in a post-Cold War context, to cling to the status quo in its military relationship with the United States as well as with cementing institutional ties with the European Union through its European Economic Area (EEA) membership and later its participation in the Schengen border control project.[34] In 1994, the Americans seriously contemplated removing their remaining fighter jets from Iceland. It would have practically ended their military presence, undermining the long-standing Icelandic argument that the Defense Agreement was not only about broader U.S. and NATO interests but also about the defense of Iceland. Still, after facing stiff Icelandic resistance, they refrained from doing so for the sake of the bilateral relationship. As for Iceland's inclusion in the European project, its participation in the EEA amounted to an associate EU membership because it relied on the "four freedoms," underpinning the European single market: the free movement of goods, persons, services, and capital.[35] Yet it exempted Iceland from participation in the EU's Common Fisheries Policy and Agricultural Policy, allowing it keep full control over its fishing grounds.

Despite prioritizing its relationship with the United States and the European Union, Iceland took part in several Northern regional intergovernmental initiatives during this period. Already in the late 1980s, Iceland had been part of discussions to establish the International Arctic Science Committee (IASC), which came into being in 1990 as well as an intergovernmental forum on Arctic issues. The initiative led to the adoption of the Arctic Environmental Protection Strategy (AEPS) in 1991 and to the formation of the Arctic Council (AC) in 1996. Still, it was not clear whether Iceland qualified as an Arctic state and whether it should be included in this process. Informal efforts to establish Arctic scientific cooperation had already started in 1986, with U.S., Norwegian, and Canadian scientists taking the lead. The idea was that the so-called Arctic "rim states"—Canada, the Soviet Union, the United States, Norway, and the Kingdom of Denmark in respect to Greenland—which had territories bordering the Arctic Ocean would form the nucleus of such cooperation. The Swedes and Finns argued, however, that they should be involved because they had territories above the Arctic Circle. The Icelanders made the same claim on the grounds that the Arctic Circle passed through Grímsey Island, which lies close to the Icelandic North coast and is part of its sovereign territory. In addition, what bolstered Iceland's case was that its Exclusive Economic Zone (EEZ) extended into the Arctic Greenland Sea, whereas Sweden and Finland had no Arctic coastlines.

The Soviets were prepared to accept that Finland and Sweden be defined as Arctic states, but since Iceland's territory north of the Arctic Circle was so small, they opposed granting it such a status. While they did not oppose the participation of the Swedes, Finns, and Icelanders in preparatory meetings on establishing IASC in the late 1980s, they continued their efforts to limit the participating states to those having a privileged interest in the Arctic, especially, the Arctic rim states. Indeed, at first, the format of these meetings was dubbed 5+3. One immediate effect, however, of Gorbachev's Murmansk speech was to soften Soviet Arctic policies, which had been based on cultivating bilateral relationships and restricted to the "rim states," and to pave the way for more extensive transnational and intergovernmental cooperation.[36]

While the Soviets continued to express reservations about Iceland's place in Arctic cooperation,[37] its inclusion was supported by the other Nordic countries, which wanted to have a strong Nordic voice in Arctic institutional venues. In the end, the Soviets relented and agreed to a definition of Arctic states comprising the eight states, a format that formed the basis of IASC, the AEPS, and, later, the Arctic Council. Thus, Iceland was involved in Arctic political cooperation and scientific research activities from the start. The 5+3 debate was—as we will see—revived in the late 2000s, when the five Arctic "rim" or "coastal" states began to meet separately on the grounds that they had special interests due to their jurisdictional claims in the Arctic Ocean. Yet, at the beginning of the 1990s, intergovernmental initiatives designed to advance Arctic cooperation at the subregional level were inclusive. The Arctic Environmental Protection Strategy was the first multilateral accord on pollution and environmental protection in the region and composed of four working groups: the Arctic Monitoring and Assessment Programme (AMAP), Conservation of Arctic Flora and Fauna (CAFF), Emergency Prevention, Preparedness and

Response (EPPR), and Protection of the Arctic Marine Environment (PAME).[38]

Iceland also became a founding member of the Barents Euro-Arctic Council (BEAC) in 1993.[39] It was rooted in a Norwegian post-Cold War initiative on the Barents Euro-Arctic Region (BAER) aimed at stabilizing its relationship with Russia—with the support of Sweden and Finland—to reduce military activities, combat environmental threats, and address "under-development" in the region. In addition, to these four core countries, the EU, Denmark, and Iceland were added to give it a broader intergovernmental profile.[40] From the start, Iceland had a minuscule stake in the BEAC. But since it was a forum for discussing economic and environmental problems, the Icelandic government wanted to promote its own maritime interests and contribute to the "modernization" of the region's fishing industry.[41] After the Arctic Council was established, Iceland sought to enhance cooperation between it and the BEAC on common projects,[42] but it was never formalized. Another venue that Iceland took part in was the Conference of Parliamentarians of the Arctic Region, which was established in 1993, representing the eight Arctic countries and the European Parliament; it also included Indigenous peoples as permanent participants as well as external observers. The Icelandic government supported the initiative, and the first Conference of Parliamentarians was held in Reykjavík. The Standing Committee of Parliamentarians of the Arctic Region (SCPAR), which became operational in 1994, was entrusted with responsibility for the activities of this forum between conferences.[43]

In Iceland, there was, as noted, cross-political support for Canada's efforts in the early 1990s to establish an Arctic intergovernmental organization[44] with a robust organizational structure, a broad-based mandate, and an agenda that included military issues.[45] Prior to the establishment of the Arctic Council, some Icelandic ministers were critical of U.S. attempts, which ultimately proved successful, to dilute its power by rejecting the idea of a permanent secretariat and by excluding military topics from its discussion.[46] After the establishment of the AC in 1996, which included the eight Arctic states and permanent participants from Indigenous groups and later extra-regional observers, Iceland agreed to host two of the AC's working group secretariats, PAME and CAFF, which were incorporated into the Council from AEPS. The soft-law structure of the Arctic Council made it easier not only to include the permanent participants but also to allow the observers—non-Arctic states, intergovernmental and interparliamentary organizations, as well as international non-governmental organizations—to take part in its activities.

Yet Icelandic Foreign Minister Halldór Ásgrímsson continued to voice misgivings, publicly, about the AC's lack of a secretariat; he also regretted that SCPAR had not been granted an observer status in the Council under the Ottawa Declaration on the formation of the AC.[47] Moreover, he was at pains to make the point that environmental challenges facing the region should not deter Arctic resource utilization, and with a subtextual reference to Iceland's dependence on fisheries, that this included "marine resources."[48] In 1996 and 1997, Iceland fought a lonely and losing battle to prevent the powerful World Wildlife Fund (WWF)[49] from becoming a permanent AC observer because of its opposition to whaling and because of its controversial claims about the state of cod stocks in the North Atlantic,[50] which had been followed up by a consumer boycott campaign against fish products.[51] Still, the admission of the WWF as an observer to the Arctic Council in 1998 was supported by the other Arctic states, which Iceland acquiesced to. Another example of Iceland's preoccupation with protecting its domestic fishing interests was that it put much emphasis—in the AC's first years—on making sure that the Council and other North Atlantic and Arctic institutions would not interfere with the sovereign rights of Arctic states to regulate, on their own, the exploitation of their natural resources.[52] Iceland also became a part of the Northern Dimension (ND) policy—together with the EU, Russia, and Norway—when it was launched in 1999, even if the geographical boundaries of this cooperation venue were mostly limited to the Baltic Sea and Barents Region.[53] The aim was to create a platform—which had been heavily promoted by Finland as part of its accession to the European Union in 1995—for the EU-Russia relationship in the fields of the economy, security, science, and culture. The participation of Norway and Iceland was mostly a symbolic way to complement this cooperation.[54]

Such Northern subregional initiatives served the purpose of promoting specific political and economic ends, especially the facilitation of Russia's integration into Western structures in a post-Cold War environment. For Iceland, membership in these institutional arrangements raised its international profile at minimal costs. It also involved participation in transnational networks that could have a surface political value as well as provide business opportunities and enhance educational cooperation. Ideally, the dynamics of regionalism have the potential to fill a region with substance, such as economic interdependence, institutional ties, political content, and in some cases, cultural belonging. When this process has advanced far enough for the region to attain some intrinsic common features, it has been described in terms of "regionness."[55] For Iceland, however, the Arctic Council became the only one of the subregional organizations established in the 1990s, which could be said to have partly fulfilled the goal of combining "national interests" with regional identities. The BEAC cooperation and the Northern Dimension were common interest-based platforms, but given that their focus was on the Barents Sea and the Baltic Sea, geographic distance meant that their activities had limited effect on Iceland.

In addition to joining subregional organizations, Icelandic governments promoted a scientific agenda on the Arctic during this period. The Stefansson Arctic Institute (SAI) was established in 1998 in Akureyri in northeastern Iceland under the auspices of the Icelandic Ministry for the Environment and Natural Resources to promote multi-disciplinary research and sustainable development in the Arctic. It was named after the famous, if controversial, Canadian-Icelandic explorer Vilhjalmur Stefansson,[56] who had visualized a polar Mediterranean, with military naval activities and global trading networks, and who had predicted that such developments would shape the political and military future of the world.[57] Ólafur Ragnar Grímsson, who became President of Iceland in 1996, was also an early champion of the Arctic cause. Before taking office, he had been a feisty politician and a former Finance Minister who had been a member of three parties during his political career: the centrist, rural Progressive Party, a short-lived Social Democratic splinter party, and the left-wing Popular Alliance—the forerunner of the Left Green Movement—which he headed from 1987 to 1995. During the early years of his presidency,[58] he took a leading part in initiating together with the Finnish Arctic scholar Lassi Heininen the Northern Research Forum (NRF). It was established in 1999 as an international platform, with participants from countries in the North, for a dialogue between members of the research community and other stakeholders, such as politicians, business leaders, civil servants, community leaders, and NGO representatives. It organized conferences in different countries based on the model of a "village square" where scholars, students, and policymakers would come together to discuss the post-Cold War changes that were taking place in the North.[59]

Even if it paid more attention to the Arctic—especially in the field of science cooperation—Iceland did not prioritize the region in its foreign policy. Whereas interaction increased among Arctic states and Arctic

Indigenous peoples, partly through their elevation to the status of permanent participants in the Arctic Council, the region did not get much traction in international politics.[60] The agenda was dominated by other issues, such as the adjustment to a post-Cold War U.S. unipolar order, economic globalization, the integration of former East bloc states to Western institutional structures, and civil and ethnic conflicts in the former Yugoslavia and Rwanda. Nonetheless, when Iceland chaired the Arctic Council from 2002 to 2004, the first indications of the impact of climate change on the Arctic were becoming known. During its tenure, the report *Arctic Climate Impact Assessment* (ACIA)—the first comprehensive study on the subject—was published. Reflecting a broad interdisciplinary approach, it showed how changes in global temperatures affected human, health, social cultural, economic conditions in the region.[61] Iceland used its position at the helm of the AC to stress Arctic socio-economic developments, which were also emerging as an important topic, to support the rights of Arctic Indigenous peoples[62] after the publication of the *Arctic Human Development Report*,[63] and to promote increased access to information technologies in the region.[64]

Even if Iceland's stewardship of the AC coincided with seminal developments relating to climate change, its policies did not focus on mitigating this trend or on reducing emissions. Instead, the Icelandic government, initially, saw global warming as an opportunity to exploit the Arctic region for economic benefit. As if in a response to Gorbachev's promise of opening of the Northern Sea Route, the Icelandic Foreign Ministry created a working group to study the material potential of prospective Arctic sea lanes. It also published a report entitled *North Meets North: Navigation and the Future of the Arctic*, in which a new nationalist Arctic frontier narrative was forged and articulated. The report blended romanticized historical accounts of Arctic navigation and exploration with scientific discussion of contemporary Arctic shipping and the possibilities offered by the future shortening of transport routes between Asia and Europe. It made the point that in the Middle Ages, "the Norse Vikings"—or the "nomads of the Sea"—had cruised "the North Atlantic in their longships, seeking fame and fortune" and established "settlements in new and unheard-of lands far to the North."[65] From Iceland, they had launched expeditions further into the unknown and settled down for a period of time in Greenland and on the continent of North America. After the Viking period, the voyages also came to an end—and, for centuries, Iceland had become dependent on other countries for international trade.

But now—the argument went—when the old dream of establishing new sea routes due to the melting of Arctic ice cover was slowly coming true, Iceland was, again, poised to play a significant role in maritime commerce.

Judged by its North Atlantic location mid-way between North Europe and the East Coast of North America, Iceland was seen as ideal for a transshipment port, which could equally serve as a hub for sea transport across the North Atlantic and the Arctic Ocean passage when it opened.[66] To promote this narrative, the Icelandic Foreign Minister, Valgerður Sverrisdóttir, went so far as to portray Iceland's future role as a transshipment center in terms of a green initiative, which, in an age of climate change, was not only good for the world economy but also the environment. Since Iceland had begun experimenting with hydrogen technology in public transport, she added that the next step would be to apply the new technology to marine transportation.[67] In addition, the phrase "Arctic Mediterranean"—coined by the explorer Vilhjalmur Stefansson in the early twentieth century—was resuscitated to highlight materialist potentials based on prospective new sea lanes and transarctic trade.[68] Icelandic officials appropriated, reformulated, and repackaged Stefansson's vision of all-year commercial sea routes around the Arctic, with ports, naval bases, and weather stations on strategically placed islands as part of an effort to highlight Iceland's place in the region. Thus, with changes in global trade patterns coupled with the linking of the North Atlantic with the Pacific, Iceland was on the cusp of becoming a key transarctic site.

Such highly inflated rhetorical attempts to draw attention to the Arctic as a source of future riches and to "greenwash" economic activities in the region lacked credibility. Despite ice-melting—with summer navigation possible within decades—it was not predicted that the Arctic sea routes would be open and readily accessible in winter. Indeed, what was missing in these speculations about new trade routes and transarctic navigation was that the Arctic Ocean was to remain ice-covered for the larger part of the year. Still, much was made of the crossing over the North Pole between the Bering Strait and the Fram Strait (between Greenland and Svalbard), which was the shortest route between the North Atlantic and North Pacific Oceans. If readily navigable, it was optimistically asserted that this route could shorten transport distances between Far East and European ports by 40% and become economically attractive as an alternative to global maritime trade routes, using the Suez and Panama canals.[69] While transarctic shipping was not yet projected by Icelandic politicians as a substitute for current transportation routes, the argument was made that it was bound to supplement them by providing more capacity for increased transportation volume.[70]

There were surely no illusions about the political influence of the Arctic Council. Gunnar Pálsson, the chair of the Senior Arctic Officials (SAO) when Iceland headed the AC, compared the organization to an "amateur football club," with "no permanent secretariat, no set logo, no regular budget."[71] He added that it was as an accountable, decentralized, and democratic forum, constituting an "intricate web of interactions with regional authorities, parliamentarians, civil society and scientific experts."[72] Instead of being manipulated by a superlayer of international civil servants, decisions were taken and followed up on by sovereign governments. Yet, despite being a circumpolar voice, he argued that it lacked international profile as a regional body.[73] In the Foreign Minister's annual foreign policy report to the Icelandic parliament in 2004, there was only a brief passage on Iceland's chairmanship of the Arctic Council apart from a discussion of the potential of the Northern Sea Route. Rather than turning its attention on the "North," the Center-Right government—made up of the conservative Independence Party and the centrist Progressive Party, which had been in power since 1995—continued to emphasize Iceland's claim to an extended continental shelf in the "South." At stake was the Hatton Rockall Area around an uninhabited island in the North Atlantic Ocean, which was also claimed by the United Kingdom, Ireland, and Denmark, based on the hope of a future exploitation of oil and gas.[74] This issue had been pushed almost single-handedly and doggedly by an influential conservative parliamentarian, Eyjólfur Konráð Jónsson—who had been the chair of Iceland's Foreign Relations Committee in the late 1980s and early 1990s.[75] It became a core part of Iceland's claims beyond its 200-mile EEZ—and it was most recently reiterated in 2021—even if it continues to contested by the above-mentioned states.[76]

In contrast, when Icelandic government officials discussed the "North," it was not to make jurisdictional demands but to protect distant waters fishing interests, especially in the so-called Loophole in the central Barents Sea and in the waters around Svalbard.[77] Iceland, had, as noted, never accepted the legality of Norway's "Fishery Protection Zone," which had been established in 1977, despite recognizing its sovereignty over Svalbard. When the Norwegians began, in the 1990s, arresting Icelandic fishing vessels without quotas or historical fishing rights in the Svalbard area, it led to a serious diplomatic dispute.[78] Iceland threatened to refer Norway's enforcement policies in the "Fishery Protection Zone" to the International

Court of Justice (ICJ).[79] When the two sides—together with Russia—reached an agreement in 1999, which granted Iceland a fishing quota in the Barents Sea in exchange for a quota in Iceland's EEZ, it removed the incentive to do so.

The international regime for the Svalbard archipelago addresses a spatially defined area, which used to be a terra nullius or a territory belonging to no sovereign state. While putting Svalbard under Norwegian sovereignty, the Spitsbergen Treaty subjected Norway to a number of conditions regarding the equitable rights of other treaty signatories to ensure development and peaceful utilization of the archipelago. Further, the treaty does not cover the continental shelf, which was an unknown concept in international law when it was signed. This could spell trouble for future governance and regional stability if those signatories that reject the Norwegian legal position that Svalbard does not generate a continental shelf of its own decide to contest it, for example, through legal proceedings. During the Cold War, the British and, to a lesser degree, the Americans only mentioned their disagreement with the Norwegian position in times of political stability in Western-Soviet relations.[80] When things became more contentious, they backed off in a show of solidarity with a fellow NATO ally.[81] The Svalbard issue was briefly revived by the British in the 2000s to reiterate the point that their legal position remained unchanged, but it did not lead to a major challenge against Norwegian sovereign rights over the archipelago.[82]

To be sure, the fishing dispute with Iceland flared up again in the mid-2000s, when the Norwegian government restricted Icelandic herring fishing within the "Fishery Protection Zone." In 2004, the Icelandic government announced that it was beginning preparations for proceedings against Norway before the ICJ[83] in response to what it termed "repeated violations" of the Spitsbergen Treaty by its decision to limit herring fishing of other states around the island.[84] The move irritated the Norwegians who have always acted forcefully when Norway's sovereignty over the Svalbard archipelago has been challenged.[85] The strong feelings of the Norwegians for Svalbard were in line with their use of the term "High North." The concept reflected their instrumental use of language to define geographic space in political terms with the aim of merging a national Norwegian narrative, focusing on the European North with broader Arctic perspectives.[86] What the Norwegians were concerned with was, among other things, to gain international support for their position on Svalbard, especially the "Fishery Protection Zone," which was contested by other countries than Iceland, such as Spain and Russia, to reduce the risk of confrontation in the Arctic because of overlapping territorial claims and to manage their relations with Russia.[87] As was the case in the 1990s, Iceland did not act on its ICJ threat. Any legal proceedings would be time-consuming and costly, and it would take the court years before issuing a ruling.[88] Thus, Iceland refrained, in the end, from pushing an issue that would have caused a major political rupture in its relationship with Norway.

2.3 Colonial and Geopolitical Arctic Imaginaries

The phrase "Scramble for the Arctic" was used, in 2007–2008, by the global media to conjure up—in exaggerated terms—potential conflict scenarios as well as to underscore the need for an international order in the Arctic.[89] True, the Arctic Five had promised at the 2008 Ilulissat meeting to settle Arctic disputes by peaceful means. But their joint commitment to a cooperation discourse on the Arctic did not excise the deep-rooted competitive geopolitical tropes in their Arctic national narratives. When, in August 2007, two Russian submersibles reached the seabed more than two and a half miles beneath the North Pole to place the one-meter-high titanium Russian flag on the underwater Lomonosov ridge—which Moscow claimed was directly connected to its continental shelf—one Russian spokesman compared the feat to the U.S. moon landing four decades earlier. The expedition's leader, Artur Chilingarov, the veteran polar explorer and later parliamentarian, belabored the point by describing it as a "soft landing,"[90] adding that the Arctic had "always been Russian and will always be Russian."[91]

Leaving the hyperbole aside, this act not only triggered memories of superpower competition and the colonization of space, which underpinned frontier narratives during the Cold War. It also referred to a domestic political tradition: In Soviet times, technological advances followed by the establishment of polar research stations functioned as expressions of modernity as well as patriotism. The "Red Arctic myth" centered on socialism's ability to master the Arctic's natural environment and to overcome the capitalist adversary.[92] The symbolic value of Arctic engagement was maximized through military parades and the elevation of polar explorers as national heroes and through the projection of the Northern Sea Route as a source of economic wealth.[93] Chilingarov himself embodied and symbolized this reified link with the past: Having been awarded the title of "Hero of the Soviet Union in 1986," he became the "Hero of the Russian Federation" in 2008. As the last borderland for both the Soviet Union and Russia, the Arctic was, thus, instrumentalized to transmit a picture of national greatness and geographic exceptionalism to domestic and international audiences.

To be sure, after the break-up of the Soviet Union, the region had been demoted and ceased to play such an edifying role in Russian national narratives, with the Yeltsin government oscillating between integrating with the West and balancing it in the international system. Still, under Putin, the Arctic was apotheosized, again, as part of a break in Russia's post-Cold War policy—a departure from Yeltsin's ambivalence in favor of reasserting Russian sovereign power on the international stage and of being prepared to challenge the West.[94] Thus, the Arctic became intrinsic to Putin's drive to recentralize political power, to restore Russia's material wealth, and to link national interests to its historical role as a Great Power after a period of neglect and devolution. While Russia did not deny that climate change was occurring in the Arctic, its policy was—and still is—squarely focused on exploitation as it was during Soviet times. To Putin, the "mastering of the Arctic"—which included consolidating territorial claims in offshore Arctic areas—was about enhancing Russia's status-seeking abroad and about making the region the engine of its economic growth.[95] In other words, the sense of nostalgia for the Soviet era in Russia's Arctic discourse has helped legitimize its current interest in the region.

Echoing the reification of the past in the present, Scott Borgerson—in a much-noted *Foreign Affairs* article in 2008—called on the U.S. political elite to abandon its neglect of the Arctic and to assume a global leadership role to prevent the region from erupting in an armed competition for its resources. He used the nostalgic paraphrase "Go North, Young Man" as a way of drawing, in gendered terms, on a specifically male-bonding American-settler experience.[96] Even if the Arctic has never served an important role in American national identity projections, this historical reference could be interpreted as a play on the spirit of "daring and youthful masculine excitement" associated with the participation in a frontier-pioneering project based on conquest—or on desires to colonize a virgin space. Thus, the Arctic was portrayed in the same way as the colonization of the American West in the nineteenth century: as an aspiring and noble cause for young men. In its contemporary reworking, there was, predictably, no mentioning of the four hundred thousand officially classified Indigenous people who lived in the Arctic. Yet this evocation of the past was a stark reminder of the continued relevance of colonial narratives in the imagination of hawkish realist commentators on the Arctic.[97]

In the first decade of the twenty-first century, Canadian Arctic narratives were also under the spell of mythologies, identity politics, and power politics. For one thing, they centered on Canadian "true North" identities—as a way of demarcation and distance from the United States. Until the Trump Administration decided to upend a diplomatic truce, the Canadians and the Americans had "agreed to disagree" on the Northwest Passage, which Canada claims as its own but the United States sees as an international waterway like the NSR.[98] To Canada, this fabled route had been a historical and an emotional source of frontier myths. In the nineteenth century, a famous British Arctic voyage of Arctic exploration—the Franklin expedition—which had been assigned to sail through the last unnavigated sections of the passage, ended in death and destruction because the crew, who saw themselves as products of the culmination of Victorian civilization, were too proud to ask the Inuit for help. In the past decades, the Canadian government has sought to cooperate with the Inuit and using them as "territorial guardians" in the Canadian Arctic. Canada's territorial discourse has sometimes bordered on "sovereignty fetishism" as exemplified by Prime Minister Steven Harper's 2007 "use-it-or-lose-it" remark about Canadian control over the Northwest Passage.[99] It was partly meant to keep Canada at arms lengths from the United States when it came to the Arctic while, simultaneously, expressing suspicions of Russia, which had the most to gain from Arctic riches.

The Danes were also grappling with colonial and Cold War legacies in Greenland. The key question was whether Danish policies toward Greenland had been motivated, historically, less by a desire to ensure the social and economic well-being of Greenlanders than by attempts to restore an alternative form of colonial relationship within the temporal contexts of the Cold War, decolonization, and Arctic geopolitics.[100] There has long been a specific imperial formation of Denmark's involvement in Greenland marked by colonial and neo-colonial aspects, while the official line has stressed modernization, integration, and interdependence. The project used to be based on three pillars: the "moral" duty to assist Greenlanders in gaining access to social welfare, liberal economy, and technological infrastructure; the explicit wish on the part of Greenland's political and cultural elites to acquire assistance from Denmark; and Danish interest in retaining sovereignty over Greenland.[101] Irrespective of the outcome of Greenlandic independence aspirations, the tension between a Western modernization discourse and Indigenous emancipatory impulses still characterizes the Danish-Greenlandic relationship.

The Norwegian infatuation with Svalbard—and its exalted place in Norway's historical cultural imagination—is another example of how the idealization of the "High North" is mixed with territorial ambition and power politics. The Norwegians have consistently worked against attempts to delegitimize their historical interpretation of Norway's Svalbard territorial rights. Thus, they adopted an Arctic resource management approach whose aim was to achieve a balance between petroleum development, growth in maritime transport, the utilization of living marine resources, and the need for environmental protection under the banner of its 2006 High North strategy.[102] It offered a "flexible" interpretation of the High North as encompassing both the North Atlantic and the Arctic and covering, geographically, the area stretching from the Barents Sea to the Greenland Sea. From a political perspective, however, it betrayed an ambitious agenda, involving Norway's relations with neighboring states, such as Sweden, Finland, and Russia; Nordic cooperation; the relationship with the United States and Canada through the Arctic Council; and the ties with the European Union as part of the Northern Dimension (with Iceland and Russia).[103] In short, it reflected a catch-all approach, combining diverse national interests and transnational processes, such as environmental, fishing, and legal concerns with geopolitics and the exploitation of natural resources.

Irrespective of the flurry of geopolitical and propaganda activities undertaken by other stakeholders in the Arctic in the first decade of the twenty-first century, the United States refrained from prioritizing the region. It did not beef up its military presence there until after the 2014 Ukrainian conflict, even if it reaffirmed its territorial interests (Alaska) and its support for the U.S. Senate's ratification of UNCLOS. The George W. Bush Administration—preoccupied with wars in Iraq and Afghanistan—believed that it was a waste of money and resources to engage in Arctic sovereignty controls. The catchphrase was "scientific timeline"—to know when to intervene and to start investing in the Arctic at some unspecified future date in the 2030s or 2040s, when the effects of climate change such as the opening of new sea lanes would be felt.[104] In other words, the United States wanted to decide when to engage with the Arctic region depending on its own interests. By joining the four other Arctic littoral states in its support for the 2008 Ilulissat Declaration on the Arctic, it wanted to uphold the status quo in the region.[105]

While the Arctic Five saw the Ilulissat meeting as an important contribution to stability, predictability, and security in the Arctic, the exclusion of the other three Arctic states—Iceland, Finland, and Sweden—from the venue prompted adverse reactions on their part. The Icelandic government was especially vocal in its protests.[106] Representatives of Arctic Indigenous people were also critical of the Arctic Five for being sidelined. The main charge leveled by the Arctic three and some Indigenous communities was that the Ilulissat initiative undermined the Arctic Council and its legitimacy as the primary Arctic organization. It was seen as the first attempt to institutionalize a new Arctic decision-making body since the late 1980s, when efforts to cement intergovernmental political and science collaboration were initially to be limited to the Arctic "rim states." This criticism was rejected by the Arctic Five on the grounds that the Ilulissat Declaration underscored their commitment to the Law of the Sea and international law. The purpose was, however, also to reiterate and project sovereign rights and to resist any attempts—by external bodies such as the European Parliament—to internationalize the Arctic either through an Arctic treaty, like the Antarctic Treaty, or through a moratorium on the exploitation of Arctic natural resources.[107] Thus, the message was clear: that Arctic governance should be left to those states that had a direct territorial stake-holding role in the region.

2.4 Reframing a Northern Discourse: The U.S.-Icelandic Military Rupture

Even if the Arctic had been registered in Iceland's foreign policy in the 1990s and early 2000s, it was dwarfed as an issue by the crisis in the defense relationship with the United States, which stemmed from the rapidly dwindling American military interest in the country after the end of the Cold War. Having abandoned its framing of Russia as an adversary, the Bush Administration decided, in late 2002, to pull out its remaining F-15 fighter jets in Iceland and to turn the military base into a standby one. The focus on the Middle East and the "War on Terror" only made the United States more determined to reduce, drastically, its military posture in Iceland. It meant that the Americans were unwilling to reward political loyalties by establishing a link between a continued U.S. military presence and the Icelandic government's backing for the Iraq War in the spring of 2003, which proved extremely controversial in Iceland—in the run up to parliamentary elections. Yet, after encountering much Icelandic opposition, they put their withdrawal plans on hold in June 2003 to placate the Icelanders and especially its conservative Prime Minister Davíð Oddsson, who despite his pro-American orientation had reacted by threatening to abrogate the U.S.-Icelandic Defense Agreement if the jets were removed unilaterally.[108]

A diplomatic stand-off between the United States and Iceland followed, which lasted three years, or until the Bush Administration—with Defense Secretary Donald Rumsfeld playing an instrumental role—decided to ignore Icelandic demands for "minimum" territorial defense and to close down the base in the fall of 2006.[109] While some Pentagon officials opposed the move neither the Navy nor the Air Force wanted to pay for the $250 million annual costs of operating the base. The curt way the decision to put an end to a long-standing historical relationship was communicated to the Icelandic government—through a single phone call from a high-ranking State Department official to Foreign Minister Geir Haarde—deeply offended Icelandic political elites, especially those on the Right who had supported the U.S. base against considerable domestic political opposition since the early 1950s. By then, Oddsson had left the government and his successors, Prime Minister Halldór Ásgrímsson and Haarde, were not prepared to act on his threat and abrogate the Defense Agreement. This meant that the Americans had continued access to Iceland, even if no permanent troops were stationed there in case the military situation in the North Atlantic underwent changes.

The U.S. decision to leave Iceland was made before the media hype surrounding the Arctic began in earnest and the debate over global warming spilled over into geopolitics. Indeed, such topics as climate change, ice-melting, access to Arctic natural resources, territorial claims under the Law of the Sea, and the prospect of new sea lines of communication (SLOC) in the North—with their potential military ramifications—did not come up in the unsuccessful U.S.-Icelandic negotiations on the future of the Keflavík base in 2005–2006. It was only after the base closure that the Icelandic government began to rethink its options. In the post-American transition phase—from 2006 to 2008—no new strategic destination was spelled out or provisions made for alternative permanent territorial defense arrangements by approaching other Western countries. Nor were any attempts made to carve out a national security identity of an unarmed state exclusively in terms of civilian participation in international operations sponsored by multilateral organizations such as the UN. Yet the U.S. departure prompted a reevaluation of Icelandic foreign and security policy to break out of a stalemate—of what the Americans saw as an outdated dependency relationship but pro-Western Icelanders as an example of the non-reliability of the United States, when facing other geopolitical priorities. The socialist Left, which had fought the U.S. military for decades, felt vindicated, arguing that the base had always served U.S., not Icelandic, interests.[110]

After the departure of the Americans, some Icelandic officials felt that one task was to prevent a "military vacuum" from opening in the North Atlantic.[111] While there was no chance that such a scenario would develop at that time, the Center-Right government in power decided to strengthen Iceland's ties with NATO by asking for "air policing" protection to hedge its bets. The Americans had, in fact, promised to support the Icelanders if they turned to NATO in the diplomatic row over the fighter jets, but the latter had rejected it on the grounds that it was the responsibility of the United States to provide for Iceland's defense. NATO agreed to provide Iceland with "air policing light"—with individual member states assuming, on a rotational basis, air surveillance with fighter jets for several weeks a year.[112] It was a scaled-down version of the all-year round air policing conducted by NATO states in the Baltic states. The Icelandic government's emphasis on territory reflected a security policy approach that was driven by—and contingent on—state-based threat scenarios. It was not, however, devoid of references to transnational or societal security factors.[113] "Emerging" or "new" threats were part of a domestic political discourse and "soft security" cooperation with other states in the fields of maritime security and search and rescue (SAR) in the North as well as the fight against organized crime and terrorism. Thus, security cooperation arrangements were negotiated with Norway,[114] Denmark[115] and, later, with Britain,[116] and Canada.[117] The agreements were mostly symbolic, for they did not involve any obligations, but they served the political purpose of tying Iceland closer to its Western and Nordic partners on security issues.

When a coalition government between the two largest parties in Iceland, the right-wing Independence Party, and the Social Democratic Alliance, was formed in the spring of 2007, there was no mention of the Arctic in its platform.[118] At the end of the year, Icelandic Foreign Minister Ingibjörg Sólrún Gísladóttir stated for the first time that the region had become a core issue in Iceland's foreign policy,[119] but no specific Arctic initiative was launched to underpin this announcement. With Russia's resumption, in 2007, of bomber flights near Iceland—which had been abandoned following the break-up of the Soviet Union—the Icelandic case for increased NATO involvement in its defenses strengthened. The decision to engage in long-range aviation was not only motivated by a desire to restore Russia's military prowess—at home and abroad—but also to underscore its geostrategic interests in places such as the Arctic. Since the Russian Tupolev bombers did not violate Iceland's airspace, the flights were not viewed by the Icelanders as a military threat. Hence, politicians refrained from securitizing them. There was, however, uneasiness about them because the bombers occasionally engaged in politically symbolic theatrics by encircling Iceland close to its airspace.[120] The flights also raised concerns about civilian air traffic due to the lack of communications with pilots; it was thought at the time that the Russians had switched off their transponders or anti-collision equipment, but later it was revealed that they were not installed in the bombers.[121]

The inclusion of Iceland in the Russian bomber flight schedules led to a temporary Western military presence on the island as part of the periodic air policing missions. Indeed, the Icelandic government soon began to forge links between NATO air policing, Russian aviation, and Arctic geopolitics. Consistent with Cold War norms, the Russian flights were monitored and intercepted by NATO fighter aircraft, but only when the jets were stationed in Iceland for three times a year.[122] Yet a Risk Assessment Commission—which was appointed, at the end of 2007, by Foreign Minister Gísladóttir, and which was composed of academics and officials[123]—concluded in a 2009 report that Iceland faced no direct military threat from other states or alliances in the short or medium term.[124]

As part of an effort to highlight the strategic dimension of the Arctic, the Icelandic government also played on alliance politics. It was seen as a way of drawing NATO's attention to Iceland by urging it to pay more attention to the North in 2008–2009.[125] The policy was also under the influence of Norway's High North strategy, with its emphasis on Russia, climate change, and oil and gas extraction.[126] While cautioning against increased military presence in the region, Iceland supported the Norwegian push for a limited NATO surveillance role in the region.[127] The idea was met with skepticism by some NATO members that also resisted a proposal for using the NATO-Russia Council to discuss Arctic affairs. Canada, for example, was adamantly against it, preferring to deal with Russia in the Arctic outside a NATO framework,[128] for it wanted to keep non-Arctic states from the region. At a NATO seminar, which was held in Reykjavík in January 2009, the Secretary General of the Alliance, Jaap de Hoop

Scheffer, delivered a cautiously worded speech, which envisioned a minor soft security function for the Alliance—in the Arctic—in the fields of search and rescue and "ecological disaster" prevention without a formal decision on the issue or operational details.[129] There was, in fact, no agreement on assigning NATO a specific defense role in the Arctic or on establishing an Arctic "security community" under its auspices,[130] even though five out of eight Arctic states were Alliance members. Karl Deutsch's original definition of a "security community"—stemming from the 1950s and reworked in the 1990s by Emanuel Adler and Michael Barnett—was contingent on the development of a "transnational community" based on a common set of ideas and values and on mutually successful predictions of behavior. Through multiple transactions and interactions, including military cooperation, trade, migration, and tourism, a communal fabric was built.[131] Russia's opposition to a militarized collective security framework in the Arctic was clear from the start. But it was not alone: Canada continued to insist that regional security should be the responsibility of Arctic states, although the Arctic Council was forbidden to deal with military issues. In addition, the then non-NATO membership status of Finland and Sweden was seen as an obstacle to any allied security involvement in the region, even if the two countries had stepped up their military cooperation with Alliance. Finally, NATO's southern members did not want to expand the Alliance's area of operations to northern waters in the absence of military tensions or threats.[132] Since the Icelandic government had only been interested in a modest monitoring role for NATO in the Arctic and not pushed for a hard security agenda, the differences within the Alliance did not have any influence on its threat perceptions.[133] The Arctic was seen as a low-tension area, which served the interests of all Arctic states.

2.5 The Nordic Security Policy Dimension

The proposals in the 2009 report—put forward by former Norwegian Foreign Minister Thorvald Stoltenberg—on Nordic foreign and security policy cooperation fitted well with Iceland's "post-American" focus on the Arctic.[134] Indeed, the idea of Nordic collaboration in the security field was strongly supported by Icelandic politicians across the political spectrum, underscoring the elevated status of the Nordic project, in general. No maximalist agenda was needed, for there was much understanding for what separated the Nordics as what brought them together: the absence of Nordic supranational mutual dependency, the preference for national and state allegiances, and involvement in larger regional alliances such as NATO and/or the EU. Still, what made enhanced Nordic security cooperation politically attractive in Iceland was that it reflected idealistic self-perceptions of the Nordics guided by notions of common democratic values, social protection, and cultural traditions.

The initial suspicion that the preponderance of the Arctic dimension in the Stoltenberg report reflected a Norwegian-centered security agenda quickly gave way to the view that the deepening of Nordic collaboration in the North was a positive thing.[135] Stoltenberg's discussion of Nordic soft security options, such as Arctic maritime monitoring, responses, and surveillance, was welcomed. The Icelandic Coast Guard was already engaged in security cooperation with the Danish Navy, which performed regular maritime patrols near Greenland. The Icelandic government was also aware of the need for increased collaboration with Denmark and Norway within the framework of Arctic natural resource politics, or what Foreign Minister Össur Skarphéðinsson euphemistically termed the "energy triangle"—the space, encompassing North-East Greenland, Jan Mayen, and the Dragon Area near Iceland. Finally, the creation of a disaster response force—as suggested by Stoltenberg—based on existing national capabilities was also seen as a way of promoting Nordic cooperation in this area. Since there was a political consensus in Iceland on prioritizing the Arctic, the government knew that Nordic security partnerships tied to perceived Icelandic national interests were likely to attract domestic support.

The most ambitious idea in the Stoltenberg report was to propose a Nordic mutual defense commitment. The subsequent 2011 Nordic solidarity declaration did not entail a formal political and military security guarantee, even if it was already hedged by Stoltenberg's emphasis that it did not come in conflict with the Nordic countries' existing UN, EU, and/or NATO obligations. Contrary to what Stoltenberg wanted, the declaration did not spell out how the Nordic countries would respond if a Nordic country was subject to external attack or undue political or military pressure. Yet, by singling out potential societal risks, such as natural and human-induced disasters as well as cyber and terrorist attacks, it was consistent with Iceland's emphasis on "soft" rather than "hard" security, which was already covered by the collective defense clause—Article 5—of the North Atlantic Treaty, stipulating that an armed attack against one NATO member should be considered an attack on all.[136]

When Iceland applied, in 2009, for membership in the European Union, one benefit that was touted was the EU's societal security guarantee—as stated in the mutual solidarity declaration in case of a terrorist attack or natural disasters. To take advantage of it, Iceland would have to join the European Union, even if it already had deep institutional ties with it through the EEA and the Schengen arrangements. The Icelandic decision to halt the EU accession process[137] in 2013—after a Center-Right government came to power—gave the Nordic solidarity declaration more weight. For Iceland, it underscored Nordic security cooperation in the event of natural catastrophes. The Nordic countries had already reaffirmed their commitment to developing civilian crisis management readiness irrespective of national borders. They had also provided the bulk of the IMF's financial rescue package in response to the Icelandic banking collapse. Admittedly, what dented the gratitude of Icelanders was that the Nordic loan was, initially, made contingent on the settlement of bilateral Icesave disputes with Britain and the Netherlands. Since Britain had used anti-terrorist legislation to take over the assets of the Icelandic bank Landsbanki, this was a hard pill to swallow and created a nationalistic backlash.[138] It raised questions of whether the Nordic countries were more concerned with doing their bidding for Great Powers, such as Britain, than with committing to Nordic solidarity.[139] Still, after the two issues were decoupled, paving the way for the payout of the loan, such critical voices became mostly muted.

The Icelandic government was prepared to take part in most Nordic security schemes, including the hosting of air surveillance and military exercises, crisis management, and data exchange. There were, surely, no illusions about the limits of Nordic cooperation in the military field and about its non-applicability when it came to mutual defense. And while Iceland became a member of the Nordic Defence Cooperation (NORDEFCO), its participation had only symbolic meaning because of its non-armed status. Yet, again, as a security option, Nordic cooperation benefited from the fact that within the domestic domain, it was interpreted positively by opposing political factions, if for different reasons. To the Right, it complemented Iceland's NATO membership, but to the Left it de-emphasized a hard security connection. Thus, while the soft security dimension was clearly favored, the military implications were not seen as undermining the project.

The 2014 participation of Finland and Sweden in military exercises in connection with Norway's NATO air policing mission in Iceland proved to be a major symbolic event for Nordic security cooperation. It was, however, as much about Swedish and Finnish security identity experiments and military collaboration with Norway. The Icelandic government wanted to link the NATO air policing mission in Iceland directly to Nordic security cooperation. Yet, if a Russian bomber had entered Iceland's "military defense identification zone," only Norwegian fighters would have been sent to intercept it. This did not, however, prevent the Icelandic government from suggesting that Finland and Sweden would take part in such a NATO operation to highlight its own role in both projects—in Nordic cooperation and in the Western Alliance—covering both the North Atlantic and the Arctic.[140]

A decade after the publication of the Stoltenberg report, a former conservative Icelandic Justice Minister Björn Bjarnason, a long-standing realist commentator on military and security issues, was asked to follow up on it. The mandate of the report was defined as covering specifically climate change, hybrid threats, and cyber issues in addition to the political questions of how "multilateralism" and the "rules-based international order" should be strengthened and reformed. Bjarnason's 2020 report was general in tone, offering 14 non-binding proposals for closer Nordic cooperation without going into detail about their implementation.[141] In an addendum on the geopolitical situation, the report engaged with Bjarnason's primary interest in hard security issues, stressing Great Power rivalry in the Arctic and elevating U.S. perceptions of Chinese and Russian threats, while also recounting the more cautious views of Nordic government ministers who continued to portray the Arctic as a low-tension area.

His proposals were not put within this broader political and military context, except in one area, that is, with respect to China. As part of the report's recommendations on dealing with climate change, it was suggested that a common Nordic approach be adopted toward Chinese Arctic involvement, which should be pursued collectively within regional organizations, such as the Arctic Council, the Council of Europe, the Council of the Baltic Sea States (CBSS), the Barents Euro-Arctic Cooperation, and the Northern Dimension.[142] While such formal collusion had often been brought up and practiced within international institutions such as the UN, it had its distractors who feared that it would lead to increased divisions within Arctic institutions. The proposal itself echoed opinions expressed by U.S. officials about the global influence of China. It took up, for example, their frequent complaint about China's double standards with respect to the Law of the Sea[143]—that it made maritime claims in the South China Sea, which contradicted UNCLOS by exempting itself from restrictions imposed on "historical rights," even if it was a party to the treaty.[144] The report concluded that "such actions must be kept out of the Arctic."[145] Yet the question remained whether it was warranted to equate China's behavior in the South China Sea—where its geopolitical and security interests were obvious—with that in the Arctic region whose importance was minor in comparison.

Published during the height of the COVID-19 pandemic and influenced by attempts to find common Nordic ways to deal with the health crisis, the Bjarnason report did not receive the same media attention or political support as the Stoltenberg report did. When Bjarnason introduced it at an online Nordic Council meeting in 2021, there were calls from some Nordic colleagues to implement his proposals.[146] Still, even if the mandate was limited to non-military issues, it is unlikely that they will be used as a blueprint for Nordic security cooperation after the profound changes that have taken place in Europe's security landscape since its release. Russia's 2022 invasion of Ukraine led to the NATO membership of Finland and Sweden; Finland, Sweden, Denmark, and Norway have also decided to deepen their military cooperation by working on the idea to create a joint Nordic air force, comprising about 250 combat aircraft, to enable their respective air forces to operate together "in all circumstances."[147] Together with the paradigm change represented by the addition of two previously non-aligned Nordic countries to NATO, making all the Nordic states part of the Alliance, the formal advancement of such policy-related collaboration goes beyond the proposals on Nordic security cooperation discussed in the Stoltenberg and Bjarnason reports.

Notes

1. See Sumarliði Ísleifsson, *Í fjarska norðursins. Ísland og Grænland—viðhorfasaga í þúsund ár* [In the Distant North: Iceland and Greenland—a Thousand-Year History of External Images] (Reykjavík: Sögufélagið, 2020), 306–311.
2. See Kristín Loftsdóttir, "Racist Caricatures in Iceland in the early 20th century," in *Iceland and Images of the North*, ed. Sumarliði R. Ísleifsson with the collaboration of Daniel Chartier (Québec: Presses de L'université de Québec, 2011), 187–204.

3. See "Ísland og Grænland" [Iceland and Greenland], memorandum on the Icelandic government's reaction to arguments in favor of making a territorial claim to Greenland, n. d, 1993, 56, sögusafn utanríkisráðumeytis [Historical Records of the Ministry for Foreign Affairs], Þjóðskjalasafn Íslands [Icelandic National Archives—hereafter ÞÍ].
4. "Ísland og Grænland."
5. See Thor Whitehead, "Leiðin frá hlutleysi [The Road from Neutrality]," *Saga* 29 (1991): 63–121; Whitehead, "Hlutleysi Íslands á hverfanda hveli 1918–1945" [Iceland's Neutrality in Limbo], *Saga* 44, no. 1 (2006): 21–64; Whitehead, *Bretarnir koma*; Hannes Jónsson, "Íslensk hlutleysisstefna [Iceland's Neutrality Policy]," *Andvari* 114, no. 1 (1989): 203–224.
6. On the Cod Wars, see Guðni Th. Jóhannesson, *Troubled Waters: Cod War, Fishing Disputes, and Britain's Fight for the Freedom of the High Seas, 1948–1964* (Reykjavík: North Atlantic Fisheries History Association, 2007); Jóhannesson, "How Cod War Came: The Origins of the Anglo-Icelandic Fishing Dispute, 1958–61," *Historical Research*, 77, no. 198 (2004): 543–574; Guðmundur J. Guðmundsson, "The Cod and the Cold War." *Scandinavian Journal of History* 31, no. 2 (2006): 97–118; Valur Ingimundarson, "Interpreting Iceland's victories in the 'Cod Wars' with the United Kingdom," in *The Success of Small States in International Relations: Mice that Roar?* ed. Godfrey Baldacchino (London and New York: Routledge, 2023), 37–50; Ingimundarson, "A western cold war: the crisis in Iceland's relations with Britain, the United States, and NATO, 1971–74," *Diplomacy and Statecraft* 14, no. 4 (2003): 94–136; Ingimundarson, "Fighting the Cod Wars in the Cold War: Iceland's challenge to the Western Alliance in the 1970s," *The RUSI Journal* 148, no. 3 (2003): 88–94; Gunther Hellmann and Benjamin Herborth, "Fishing in the mild West: democratic peace and militarised interstate disputes in the transatlantic community," *Review of International Studies* 34, no. 3 (2008): 481–506; Sverrir Steinsson, "The Cod Wars: A re-analysis," *European Security* 25, no. 2 (2016): 256–275; Steinsson, "Do liberal ties pacify? A study of the Cod Wars," *Cooperation and Conflict* 53, no. 3 (2018): 339–355; Thorir Gudmundsson, "Cod War on the High Seas: Norwegian-Icelandic dispute over the 'Loophole' fishing in the Barents Sea," *Nordic Journal of International Law* 64 (1995): 557–572.
7. See Thor Whitehead, "Iceland in the Second World War," Ph.D. diss. (Oxford University, 1978); Whitehead, *Ófriður í aðsigi* [War Approaching] (Reykjavík: Almenna bókafélagið, 1980); Whitehead, *Stríð fyrir ströndum* [War Close to the Shores] (Reykjavík: Almenna bókafélagið, 1985); Whitehead, *Milli vonar og ótta* [Between Hope and Fear] (Reykjavík: Vaka-Helgafell, 1995); Whitehead. *Bretarnir koma* [The British Arrive]

(Reykjavík:Vaka-Helgafell, 1999); Gunnar M. Magnúss. *Virkið í norðri* [The Fortress in the North], I–III (Reykjavík: Bókaútgáfan Virkið, 1984); Donald F. Bittner, *The Lion and the White Falcon: Britain and Iceland in the World War II Era* (Archon Books: Hamden, Connecticut 1983); Michael T. Corgan, "Franklin D. Roosevelt and the American Occupation of Iceland," *Naval War College Review*, 45, no. 4 (1992): 34–54; Sólrún B. Jensdóttir Harðarson, "The 'Republic of Iceland' 1940–44: Anglo-American Attitudes and Influences," *Journal of Contemporary History*, 9, no. 4 (1974): 27–56; Guðmundur Hálfdanarson, "'The Beloved War' The Second World War and the Icelandic National Narrative," in *Nordic Narratives of the Second World War: National Historiographies Revisited*, ed. Henrik Stenius, Mirja Österberg, and Johan Östling (Lund: Nordic Academic Press, 2011).

8. See, for example, B. Schofield, *The Arctic Convoys* (London: Macdonald & Jane's Ltd., 1977); Richard Woodman, *Arctic Convoys 1941–1945* (London: John Murray, 1994).
9. See Ingimundarson, *The Reluctant Ally*, 180.
10. See Peter Kikkert and P. Whitney Lackenbauer, "The Militarization of the Arctic to 1990," in *The Palgrave Handbook of Arctic Policy and Politics*, 487–505.
11. Kikkert and Lackenbauer, "The Militarization of the Arctic to 1990," 498.
12. See, for example, Gunnar Gunnarsson, "Continuity and Change in Icelandic Security and Foreign Policy," *The Annals of American Academy of Political and Social Science*, 512, no. 1 (1990): 140–151; Albert Jónsson, *Iceland, NATO, and the Keflavik Base* (Reykjavík: Icelandic Commission on Security and International Affairs, 1989); Valur Ingimundarson, "Icelandic Domestic Politics and Popular Perceptions of NATO, 1949–1999," in *NATO—The First Fifty Years*, ed. Gustav Schmidt (London: Macmillan, 2001), 285–302.
13. On the Reagan-Gorbachev Reykjavík Summit, see Barbara Farnham, "Reagan and the Gorbachev Revolution: Perceiving the End of Threat," *Political Science Quarterly* 116, no. 2 (2001): 225–252; Kenneth L. Adelman, *Reagan at Reykjavik: Forty-Eight Hours That Ended the Cold War* (New York: Broadside Books, 2014); Archie Brown, "The Gorbachev revolution and the end of the Cold War," in *Cambridge History of the Cold War*, ed. Melvyn P. Leffler and Odd Arne Westad (Cambridge: Cambridge University Press, 2010), 244–266; Jonathan Hunt and David Reynolds, "Geneva, Reykjavik, Washington, and Moscow, 1985–8," in *Transcending the Cold War: Summits, Statecraft, and the Dissolution of Bipolarity in Europe, 1970–1990*, ed. Kristina Spohr and David Reynolds (Oxford: Oxford University Press, 2016), 151–79; Jack F. Matlock, *Reagan and Gorbachev: How the Cold War Ended* (New York: Random

House, 2004); Ronald Reagan, *The Reagan Diaries,* ed. Douglas Brinkley (New York: Harper Collins, 2007); George P. Shultz, *Turmoil and Triumph: My Years as Secretary of State* (New York: Charles Scribner's Sons, 1993); William Taubman, *Gorbachev: His Life and Times* (New York: W.W. Norton, 2017); Raymond L. Garthoff, *The Great Transition: American-Soviet Relations and the End of the Cold War* (Washington, D.C.: Brookings Institution, 1994).

14. See Ronald Reagan, *An American Life* (New York: Simon and Schuster, 1990), 683.
15. Mikhail Gorbachev, "The Speech in Murmansk at the ceremonial meeting on the occasion of the presentation of the Order of Lenin and the Gold Star Medal in the city of Murmansk, October 1, 1987," accessed November 16, 2022, https://www.barentsinfo.fi/docs/Gorbachev_speech.pdf.
16. On Gorbachev's Murmansk speech and its implications, see Kristian Åtland, "Mikhail Gorbachev, the Murmansk Initiative, and the Desecuritization of Interstate Relations in the Arctic," *Cooperation and Conflict* 43, no. 3 (2008): 289–311; Douglas C. North, *Governance within the Far North* (London and New York: Routledge, 2016), 13; Geir Hønneland, *Arctic Euphoria and International High North Politics* (Singapore: Springer, 2017), 8; Ronald Purver, "Arctic Security: The Murmansk Initiative," in *Soviet Foreign Policy: New Dynamics, New Themes,* ed. Ronald Purver (New York: St. Martin's Press, 1989), 182–203; Heather Exner-Pirot, "Between Militarization and Disarmament: Challenges for Arctic Security in the Twenty-First Century," in *Climate Change and Arctic Security: Searching for a Paradigm Shift,* ed. Lassi Heininen and Heather Exner-Pirot (Cham: Palgrave/Macmillan, Springer Nature: 2020), 91–206.
17. There is one exception, however; see Pavel Devyatki, "Arctic exceptionalism: a narrative of cooperation and conflict from Gorbachev to Medvedev and Putin," *Polar Journal* 13, no. 2 (2023): 336–357.
18. See Alexander Sergunin, "Russia and Arctic Security: Inward-Looking Realities," in *Breaking Through: Understanding Sovereignty in the Circumpolar Arctic,* ed. Wilfried Greaves and P. Whitney Lackenbauer (Toronto: University of Toronto Press, 2021), 117.
19. Gail Fondahl, Aileen A. Espiritu, and Aytalina Ivanova, "Russia's Arctic Regions and Policies," in *The Palgrave Handbook of Arctic Policy and Politics,* 195–216.
20. See Elizabeth Buchanan, *Red Arctic: Russian Strategy Under Putin* (Lanham, Rowman & Littlefield, 2023).
21. See "U.S. Position on Soviet Proposals for Arctic Reaction," n.d., [1988] contained in a memorandum (Helgi Ágústsson), March 30, 1988, Folder,

B/31, 1980–1998, sendiráð Íslands í Osló [Icelandic Embassy in Oslo], utanríkisráðuneytið [Iceland's Ministry for Foreign Affairs] 2011, ÞÍ.

22. See Douglas C. North, *Governance within the Far North*, 14.
23. President Reagan's written responses to questions submitted by the Finnish newspaper *Helsingin Sanomat*, The American Presidency Project, UC Santa Barbara, May 19, 1988, accessed November 16, 2023, https://www.presidency.ucsb.edu/documents/written-responses-questions-submitted-the-finnish-newspaper-helsingin-sanomat.
24. See Geir Hønneland, *Arctic Euphoria and International High North Politics*, 8.
25. See, for example, Steven G. Sawhill, "Cleaning-up the Arctic's Cold War Legacy: Nuclear Waste and Arctic Military Environmental Cooperation," *Cooperation and Conflict* 35, no. 1 (2000): 5–36.
26. "Ræða Steingríms Hermannssonar utanríkisráðherra um utanríkismál" [Speech by Foreign Minister Steingrímur Hermannsson on Foreign Affairs], *Þingtíðindi* [Icelandic parliamentary records], February 22, 1987, accessed May 23, 2023, https://www.althingi.is/altext/110/s/pdf/0596.pdf.
27. Þingræða [Parliamentary Speech] by Guðmundur H. Garðarsson, April 25, 1989, accessed May 5, 2023, *Þingtíðindi*, https://www.althingi.is/altext/111/r3/3675.html.
28. See, for example, the editorial in *Morgunblaðið*, October 4, 1987; see also þingræða fjármálaráðherra [Parliamentary Speech by Finance Minister], Jón Baldvin Hannibalsson, *Þingtíðindi*, February 26, 1988, accessed November 16, 2022, https://www.althingi.is/altext/raeda/?rnr=3418<hing=110.
29. See, for example, Elliot L. Richardson, "Jan Mayen in Perspective," *The American Journal of International Law* 82, no. 3 (1988): 443–458.
30. See report on scientific research in the Arctic and the establishment of IASC by Magnús Magnússon (1989), Folder, Norðurheimssvæði [The Arctic], B/31, 12. P.1., sendiráð Íslands í Osló, utanríkisráðuneytið 2011, ÞÍ.
31. See, for example, Áslaug Ásgeirsdóttir, *Who Gets What? Domestic Influences on International Negotiations Allocating Shared Resources* (New York: Suny Press, 2008); Áslaug Ásgeirsdóttir, "Á hafi úti: Áhrif hagsmunahópa á samninga Íslendinga og Norðmanna vegna veiða úr flökkustofnum" [In Distant Waters: The Influence of Interest Groups on Agreements between Iceland and Norway on Straddling Stocks Fishing Activities], in *Uppbrot hugmyndakerfis. Endurmótun íslenskrar utanríkisstefnu 1991–2007*[The Unravelling of an Ideational System: Reconfiguring Icelandic Foreign and Security Policy, 1991–2007], ed. Valur Ingimundarson (Reykjavík: Hið íslenska bókmenntafélag, 2008),

349–371; Bjarni Már Magnússon, "The Loophole Dispute from an Icelandic Perspective," Working Paper (Centre for Small States, University of Iceland, 2010), 1–31; Bjarni Már Magnússon, "Nokkrir farvegir fyrir hafréttardeilur Íslands við erlend ríki og alþjóðastofnanir" [Several Options in Iceland's Law of the Sea Disputes between Iceland and Other States and International Organizations], *Tímarit Lögréttu* 10, no. 1 (2014): 9–20; Arnór Snæbjörnsson, "Smugudeilan: Veiðar Íslendinga í Barentshafi 1993–1999 [The Loophole Dispute: Icelandic Fishing the Barents Seas, 1993–1999"] (MA thesis, University of Iceland, 2015); Thorir Gudmundsson, "Cod War on the High Seas: Norwegian-Icelandic dispute over the 'Loophole' fishing in the Barents Sea"; see also memorandum on the Svalbard Case (Gunnar G. Schram), October 15, 1993, Folder, B/549, bréfasafn [Letter Collection], 1965–1994, utanríkisráðuneytið 2011, ÞÍ; memorandum "Aðild Íslands að Svalbarðasamningnum" [Iceland's Accession to the Svalbard Treaty] (Helgi Gíslason), May 11, 1994, Folder, B/550, bréfasafn, 1941–1994, utanríkisráðueytið, 2011, ÞÍ.

32. See memorandum (utanríkisráðuneytið) to sendiráð Íslands í Moskvu [Icelandic Embassy in Moscow], March 22, 1994, Folder, B/542, bréfasafn, 1973–1994, Svalbarði, utanríkisráðuneytið 2011, ÞÍ.
33. See Øvind Østerud and Geir Hønneland, "Geopolitics and International Governance in the Arctic," *Arctic Review on Law and Politics* 5, no. 2 (2014): 156–176.
34. See, for example, Baldur Thorhallson and Hjalti Thor Vignisson, "A Controversial step: Membership of the EEA," in *Iceland and European Integration: On the Edge* (London: Routledge, 2004), 38–49; Baldur Thorhallsson, "Evrópustefna íslenskra stjórnvalda: Stefnumótun, átök og afleiðingar," in *Uppbrot hugmyndakerfis. Endurmótun íslenskrar utanríkisstefnu 1991–2007*, 67–136; Valur Ingimundarson, "Frá óvissu til upplausnar. 'Öryggissamfélag' Íslands og Bandaríkjanna 1991–2006 [From Uncertainty to Dissolution: The Icelandic-U.S. 'Security Community 1991–2006']," in *Uppbrot hugmyndakerfis*, 1–66; Jóhanna Jónsdóttir, *Iceland and the EU: Europeanization and the European Economic Area* (Abington, Oxon, New York: Routledge, 2013).
35. See, for example, Baldur Thorhallsson, "Iceland: A reluctant European," in *The European Union's non-members: independence under hegemony?* ed. Erik Oddvar Eriksen and John Erik Fossum (London and New York, 2015), 118–136.
36. See Louwrens Hacquebord, "How Science Organizations in the Non-Arctic Countries became Members of IASC," *IASC after 25 Years*, Special Issue of the *IASC Bulletin* (2015): 21–26; see also Odd Rogne, "Initiation of the International Arctic Science Committee (IASC)," *IASC after 25 Years*, Special Issue of the *IASC Bulletin* (2015): 9–19.

37. See Magnús Magnússon's report on scientific research in the Arctic and the establishment of IASC.
38. See Elana Wilson Rowe, *Arctic governance. Power in cross-border cooperation* (Manchester: Manchester University Press, 2018), 71.
39. BEAC members are the following: Denmark, Finland, Iceland, Norway, Russia, Sweden, and the European Commission.
40. See Hønneland, *Arctic Euphoria and International High North Politics*, 7, 25–39.
41. See "Statement, Halldór Ásgrímsson, Minister of Foreign Affairs and External Trade of Iceland," 5th Ministerial Session of the Barents Euro-Arctic Council (BEAC), n.d. [1998], B/225,1991–1999, sendiráð Íslands í Brussel [Icelandic Embassy in Brussels], utanríkisráðuneytið 2011, ÞÍ.
42. See memorandum (Ólafur Egilsson and Finnbogi Rútur Arnarson),"Co-operation and Coordination between the Barents Council and the Arctic Council, Barents Euro-Arctic Council, Committee of Seniors Officials," February 18, 1997, utanríkisráðuneytið, ÞÍ.
43. See a description of the Conference of Parliamentarians of the Arctic Region, accessed February 10, 2024, https://arcticparl.org/about/.
44. Svein Vigeland Rottem, *The Arctic Council: Between Environmental Protection and Geopolitics* (Singapore: Springer, 2020), 4–5.
45. Donat Pharand, "Draft Arctic Treaty: An Arctic Region Council" [reprint], *Northern Perspectives* 19, no. 2 (1991): n.p.; Donat Pharand, "The Case for an Arctic Region Council and a Treaty Proposal," *Revue generale de droit* 2, 23 (1992): 163–195; Piotr Graczyk and Svein Vigeland Rottem, "The Arctic Council: soft actions, hard effects?" in *Routledge Handbook of Arctic Security*, 221–233; Piotr Graczyk, "Observers in the Arctic Council—Evolution and Prospects," *Yearbook of Polar Law* 3 (2011): 594–596.
46. See *Alþýðublaðið*, March 3, 1995; see also *Morgunblaðið*, September 10, 1994; *Tíminn*, November 26, 1994.
47. SCPAR received an observer status in the Arctic Council in 1998.
48. See "Statement by Halldór Ásgrímsson, Minister for Foreign Affairs of Iceland," September 19, 1996, Folder, B/2006, 1996–1997, Norðurskautsráðið [Arctic Council], sjávarútvegsráðuneytið [Ministries of Fisheries], ÞÍ.
49. Its name was changed to World Wide Fund for Nature in 1986, but in the United States and Canada, the original name is still officially used.
50. See memorandum (Ólafur Egilsson) to sjávarútvegsráðuneytið (Tryggvi Felixson), January 31, 1997, Folder, B/2006, 1996–1997, Norðurskau tsráðið, sjávarútvegsráðuneytið, ÞÍ; see also memorandum (Magnús Jóhannesson) to Ólafur Egilsson, January 22, 1997, Folder, B/2006, 1996–1997, Norðurskautsráðið, sjávarútvegsráðuneytið, ÞÍ.

51. WWF International apologized to the Icelandic government for inaccuracies and misleading statements in an advertisement in *Time* against cod fishing in the North Atlantic, claiming that it had not targeted Iceland. See Peter Prokosch, Coordinator of WWF Arctic Program, to Ambassador Ólafur Egilsson, January 30, 1997, Folder, B/2006, 1996–1997, Norðurskautsráðið, sjávarútvegsráðuneytið, ÞÍ.
52. See, for example, memorandum (Gunnar Gunnarsson) to Guðrún Eyjólfsdóttir, April 11, 2000, Folder, B/584, 1998–2000, sjávarútvegsráðuneytið, ÞÍ.
53. Members of the Northern Dimension policy cooperation are the European Union, Russia, Norway, and Iceland.
54. See Mette Sicard Filtenborg, Stefan Gänzle, and Elisabeth Johansson, "An Alternative Theoretical Approach to EU Foreign Policy: 'Network Governance' and the Case of the Northern Dimension," *Cooperation and Conflict* 37, no. 4 (2003): 387–407.
55. See Raimo Värynen, "Regionalism: Old and New," *International Studies Review* 5, no. 1 (2003): 25–51.
56. As a pioneering explorer, Stefansson discovered major land masses, such as Brock, Borden, Meighen, and Lougheed islands. His legacy was romanticized in Iceland as underscored by the decision to name an Arctic institute after him. But in Canada, his reputation was tarnished following a disastrous 1913 expedition, which was sponsored by the Canadian government and which cost many lives. As the leader of the expedition, he was criticized for failing to prepare the crew and scientists for Arctic survival conditions and for abandoning an ice-locked ship, Karluk. Similarly, in 1922, he was blamed for another aborted expedition, involving the unauthorized attempt to colonize Wrangel Island, which belonged to Russia, and resulting in the death of crew members. See, for example, *Jennifer Niven, The Ice Master: The Doomed 1913 Voyage of the Karluk and the Miraculous Rescue of Her Survivors (New York: Hyperion, 2000).*
57. See Peter Kikkert and P. Whitney Lackenbauer, "The Militarization of the Arctic to 1990"; see also Paul Dukes, "Vilhjalmur Stefansson: The Northward Course of Empire, The Adventure of Wrangel Island, 1922–1925, and 'Universal Revolution'," *Sibirica: Interdisciplinary Journal of Siberian Studies* 17, no. 1 (2018): 1–22; Vilhjalmur Stefansson, *Discovery: The Autobiography of Vilhjalmur Stefansson* (New York: McGraw-Hill, 1964); Gisli Pálsson, *Travelling Passions: The Hidden Life of Vilhjalmur* Stefansson (Winnipeg: University of Manitoba Press, 2005).
58. About the first half of Grímsson's Presidency, see Guðjón Friðriksson's biography, *Saga af forseta* [A President's Account] (Reykjavík: Forlagið, 2008).

59. The University of Akureyri and the Stefansson Arctic Institute hosted the secretariat of the Northern Research Forum (NFS), which had a steering group composed of members from Canada, Iceland, Finland, the United States, Greenland, and Russia, and an honorary board chaired by Grímsson, which included, among others, the late Martti Ahtisaari, President of Finland; the late Lennart Meri, President of Estonia; Vaira Freiberga, President of Latvia; and Prince Albert II of Monaco. On the Northern Research Forum, see Lassi Heininen, "The Northern Research Forum—a Pioneering Model for an Open Discussion," *Arctic Circle Journal*, September 14, 2023, https://www.arcticcircle.org/journal/the-northern-research-forum.
60. See Andreas Østhagen, "The Good, the Bad, and the Ugly: Three Levels of Arctic Geopolitics," in *The Arctic and World Order*, ed. Kristina Spohr and Daniel S. Hamilton (Washington, D.C.: Foreign Policy Institute/Henry A. Kissinger Center for Global Affairs, Johns Hopkins University SAIS, 2020), 363.
61. Arctic Council, *Impacts of a Warming Arctic: Arctic Climate Change Impact Assessment* (ACIA) (Cambridge: Cambridge University Press, 2004), accessed February 10, 2024, https://www.amap.no/documents/download/1058/inline.
62. Address by Ambassador Benedikt Jónsson on behalf of Iceland's Chairmanship of the Arctic Council, at the International Round Table—Indigenous Peoples of the North and the Parliamentary System of the Russian Federation: Experience and Prospects, Moscow, March 12–13, 2003, accessed April 25, 2023, https://www.utanrikisraduneyti.is/frettaefni/ymis-erindi/nr/224.
63. Arctic Council, *Arctic Human Development Report* (AHDR) (Akureyri: Stefansson Arctic Institute, 2004), accessed November 20, 2023, file:///C:/Users/vi/Downloads/Arctic%20Human%20Development%20Report.pdf.
64. "The Icelandic Chairmanship Program," Address by Ambassador Gunnar Pálsson, Chair of Senior Arctic Officials, Northern Forum 6th General Assembly, St. Petersburg, April 24, 2003, accessed May 25, 2023, April 24, 2003, https://www.stjornarradid.is/efst-a-baugi/frettir/stokfrett/2003/04/25/Aaetlun-Islands-i-formennsku-Nordurskautsradsins-2002-2004/.
65. Report of a working group of the Icelandic Ministry for Foreign Affairs, *North Meets North: Navigation and the Future of the Arctic* [translated from the original report in Icelandic entitled *Fyrir stafni haf. Tækifæri tengd siglingum á norðurslóðum*] [Ocean Ahead: Opportunities Linked to Arctic Shipping] (Reykjavík: utanríkisráðuneytið), [February 2005], July 2006, accessed November 15, 2022, 54, https://www.stjornarradid.is/media/utanrikisraduneyti-media/media/Utgafa/vef_skyrsla.pdf.

66. *North Meets North: Navigation and the Future of the Arctic.*
67. Opening Address by Valgerður Sverrisdóttir, Minister for Foreign Affairs, in the Report *Breaking the Ice: Arctic Development and Maritime Transportation. Prospects of the Transarctic Route—Impact and Opportunities*, March 27–28 2007, 4–5.
68. See *Ísinn brotinn. Þróun norðurskautssvæðisins og sjóflutningar, horfur í siglingum á Norður-Íshafsleiðinni* [Broken Ice: Arctic Developments and Sea Transports; Prospects for Arctic Shipping] (Reykjavík: utanríkisráðuneytið, 2006), accessed May 29, 2023, https://www.stjornarradid.is/media/utanrikisraduneytimedia/media/utgafa/isinn_brotinn.pdf; "Ísland á norðurslóðum" [Iceland and the Arctic] (Reykjavík: utanríkisráðuneytið, 2009), accessed May 29, 2023, https://www.stjornarradid.is/media/utanrikisraduneyti-media/media/skyrslur/skyrslan_island_a_nordurslodum.pdf.
69. Opening Address by Valgerður Sverrisdóttir, "Breaking the Ice: Arctic Development and Maritime Transportation. Prospects of the Transarctic Route—Impact and Opportunities," 5.
70. "Breaking the Ice: Arctic Development and Maritime Transportation. Prospects of the Transarctic Route—Impact and Opportunities," 5.
71. "Arctic co-operation 12 years on: How successful?" Address by Ambassador Gunnar Pálsson, Chairman of Senior Arctic Officials Wilton Park Conference, United Kingdom, March 20, 2003, accessed March 5, 2023, https://www.utanrikisraduneyti.is/frettaefni/ymis-erindi/nr/235.
72. "Arctic co-operation 12 years on: How successful?"
73. "Arctic co-operation 12 years on: How successful?"
74. *Skýrsla Halldórs Ásgrímssonar utanríkisráðherra um utanríkismál* [Report by Foreign Minister Halldór Ásgrímsson on Foreign and International Affairs], Þingtíðindi, April 2004, accessed November 25, 2022, https://www.althingi.is/altext/130/s/pdf/1377.pdf.
75. "Eyjólfur Konráð Jónsson, 13. júní, 1928—6. Marz 1997," *Morgunblaðið*, March, 14, 1997, accessed February 3, 2024, https://timarit.is/page/1874754#page/n0/mode/2up.
76. United Nations, Division for Ocean Affairs and Law of the Sea, "Commission on the Limits of the Continental Shelf (CLCS). Outer limits of the continental shelf beyond 200 nautical miles from the baselines: Submissions to the Commission: Partial revised Submission by Iceland," April 4, 2021, accessed February 10, 2024, https://www.un.org/depts/los/clcs_new/submissions_files/submission_isl_rev2021.htm; Permanent Mission of Iceland to the United Nations, "Iceland and the United Nations," n.d., accessed February 10, 2024, https://www.government.is/diplomatic-missions/permanent-mission-of-iceland-to-the-united-nations/iceland-and-the-united-nations/.

77. Bjarni Már Magnússon, "The Loophole Dispute from an Icelandic Perspective," 11.
78. Bjarni Már Magnússon, "The Loophole Dispute from an Icelandic Perspective"; Magnússon, "Nokkrir farvegir fyrir hafréttardeilur Íslands við erlend ríki og alþjóðastofnanir"; Arnór Snæbjörnsson, "Smugudeilan: Veiðar Íslendinga í Barentshafi 1993–1999."
79. See, for example, *Morgunblaðið*, June 23, 1994.
80. Memorandum on meeting with BP on Spitsbergen, September 12, 1974, Questions on Legal Status Continental Shelf around Spitsbergen, 1974, Foreign Commonwealth Office [FCO] 33/2594, National Archives [NA], the United Kingdom (UK); J.E. Cornish to C. Hulse, 23 August 23, 1974, Questions on Legal Status Continental Shelf around Spitsbergen, 1974. FCO 33/2594, UKNA; K.B.A. Scott to C. Hulse, March 10, 1976, Legal Status of Spitsbergen, 1976. 59, FCO 33/3071, UKNA; D.A.S. Gladstone to Sir Archie Lamb, HM Ambassador, Oslo, July 18, 1979, Norway and Svalbard (Spitzbergen) FCO 33/4297, 1979, UKNA. See also U.S. Embassy Oslo to State Department, July 7, 1976, "Public Library of US Diplomacy," WikiLeaks Documents, "The 200-mile Fishery Zone and our Svalbard Policy," accessed February 2, 2024, https://wikileaks.org/plusd/cables/1976OSLO33736_b.html.
81. C. Hulse to Batstone, Legal Advisers, July 12, 1974, Questions on Legal7 Status of Continental Shelf around Spitsbergen, 1974 FCO 33/2593, UKNA.
82. See Valur Ingimundarson, "The Geopolitics of the 'Future Return': Britain's Century-Long Challenges to Norway's Control over Spitsbergen," *The International History Review* 40, no. 4 (2018): 893–915.
83. "Ræða Davíðs Oddssonar utanríkisráðherra um utanríkismál" [Address by Foreign Minister Davíð Oddsson on Foreign Affairs to the Icelandic Parliament], November 11, 2004, accessed November 25, 2023, https://www.stjornarradid.is/raduneyti/utanrikisraduneytid/utanrikisradherra/fyrri-radherrar/stok-raeda-fyrrum-radherra/2004/11/11/Raeda-Davids-Oddssonar-utanrikisradherra-um-utanrikismal/.
84. "Ræða Davíðs Oddssonar utanríkisráðherra um utanríkismál," April 29, 2005.
85. See, for example, Christopher R. Rossi, "'A Unique International Problem': The Svalbard Treaty, Equal Enjoyment, and Terra Nullius: Lessons of Territorial Temptation from History," *Washington University Global Studies Law Review* xv, no. 1 (2016): 93–136; Robin Churchill and Geir Ulfstein, "The Disputed Maritime Zones around Svalbard," in *Changes in the Arctic Environmental and the Law of the Sea*, ed. Myron H. Nordquist, Tomas H. Heidar, and John Norton Moore (Leiden:

Martinus Nijhoff, 2010), 551–593; D.H. Anderson, "The Status Under International Law of the Maritime Areas Around Svalbard," *Ocean Development & International Law* xl, no. 4 (2009): 373–84; Geir Ulfstein, *The Svalbard Treaty: From Terra Nullius to Norwegian Sovereignty* (Oslo: Scandinavian University Press, 1995); Torbjørn Pedersen, "The Svalbard Continental Shelf Controversy: Legal Disputes and Political Rivalries," *Ocean Development & International Law* xxxvii, no. 3–4 (2006): 339–358; Torbjørn Pedersen and Tore Henriksen, "Svalbard's Maritime Zones: The End of Legal Uncertainty?" *The International Journal of Marine and Coastal Law* xxiv, no. 3 (2009): 141–161; A.N. Vylegzhanin and V.K. Zilanov, *Spitsbergen: Legal Regime of Adjacent Maritime Areas*, trans. William E. Butler (Utrecht: Eleven International Publishing, 2007); Sarah Wolf, "Svalbard's Maritime Zones, their Status Under International Law and Current and Future Disputes Scenarios," Stiftung Wissenschaft und Politik, Berlin, Working Paper, FG 2, no. 2 (2013), 1–37; Valur Ingimundarson, "The Geopolitics of the 'Future Return': Britain's Century-Long Challenges to Norway's Control over Spitsbergen."

86. Jonas Gahr Støre, "Iceland and Norway—Neighbours in the High North," speech given at the University of Iceland, November 3, 2008, accessed on February 1, 2022, https://www.regjeringen.no/en/dep/ud/Whats-new/Speeches-and-articles/speeches_foreign/2008/iceland-and-norway–neighbours-in-the-hi.html?id=534706.
87. See, for example, Torbjørn Pedersen, "The Svalbard Continental Shelf Controversy: Legal Disputes and Political Rivalries," *Ocean Development & International Law*, xxxvii, no. 3–4 (2006): 339–358.
88. Arnór Snæbjörnsson, "Smugudeilan: Veiðar Íslendinga í Barentshafi 1993–1999," 74.
89. "A scramble for the Arctic: With one fifth of the world's oil and gas at stake, countries are struggling to control the once-frozen arctic," *Aljazeera*, August 12, 2010, accessed February 2, 2024, https://www.aljazeera.com/features/2010/12/8/a-scramble-for-the-arctic; "Scramble for the Arctic," *Financial Times*, August 19, 2007, accessed February 2, 2007; https://www.ft.com/content/65b9692c-4e6f-11dc-85e7-0000779fd2ac; "A mad scramble for the shrinking Arctic," *New York Times*, September 10, 2008, accessed February 2, 2024, https://www.nytimes.com/2008/09/10/opinion/10iht-edarctic.1.16040367.html.
90. "Russia plants flag on North Pole seabed," *Guardian*, August 2, 2007, accessed May 14, 2023, https://www.theguardian.com/world/2007/aug/02/russia.arctic.

91. Quoted in Nicole Bayat Grajewski, "Russia's Great Power Assertion: Status-Seeking in the Arctic," *St. Anthony's International Review* 13, no. 1 (2017): 152–153.
92. See John McCannon, *Red Arctic: Polar Exploration and the Myth of the North in the Soviet Union, 1932–1939* (Oxford: Oxford University Press, 1998).
93. Grajewski, "Russia's Great Power Assertion: Status-Seeking in the Arctic," 147–148.
94. See, for example, Allen C. Lynch, "The influence of regime type on Russian foreign policy toward 'the West,' 1992–2015," *Communist and Post-Communist Studies* 49 no. 1 (2016): 101–111; Allen C. Lynch, "The Realism of Russia's Foreign Policy," *Europe-Asia Studies*, 53, no. 1 (2001): 7–31; Rosalind Marsh, "The Nature of Russia's Identity: The Theme of 'Russia and the West' in Post-Soviet Culture," *Nationalities Papers* 35, no. 3 (2007): 555–578; Alexander Sergunin, *Explaining Russian Foreign Policy Behavior: Theory and Practice* (Stuttgart: ibidem Press, 2016); Lilia Shevtsova, *Russia: Lost in Transition. The Yeltsin & Putin Legacies (Washington, DC: Carnegie Endowment for International Peace, 2007).*
95. Grajewski, "Russia's Great Power Assertion: Status-Seeking in the Arctic," 143, 155–156.
96. See Borgerson, "Arctic Meltdown"; Borgerson, "The Great Game Moves North."
97. On masculinity and Polar ideologies, see Lisa Bloom, *Gender on the Ice: American Ideologies of Polar Expeditions* (Minneapolis: University of Minnesota Press, 1993).
98. See, for example, Donat Pharand, "The Arctic Waters and the Northwest Passage: A Final Revisit," *Ocean Development and International Law* 38 (2007): 3–69; Donat Pharand, "Canada's Sovereignty Over the Northwest Passage," *Michigan Journal of International Law* 10, no. 2 (1989): 653–678; James Kraska, "International Security and International Law in the Northwest Passage," *Vanderbilt Law Review* 42, no. 4 (2009): 1109–1132; Danita Catherine Burke, "Leading by example: Canada and its Arctic stewardship role," *International Journal of Public Policy* 13, no. 1–2 (2017): 36–52; Franklyn Griffiths, *The Politics of the Northwest Passage* (Montreal: McGill-Queens University Press, 1987).
99. See "Arctic sovereignty a priority: Harper," *CBS*, August 23, 2010, accessed February 2, 2024, https://www.cbc.ca/news/politics/arctic-sovereignty-a-priority-harper-1.951536; see also, Griffiths, "Towards a Canadian Arctic Strategy," 579–624.
100. See Thorsten Borring Olesen, "Between Facts and Fiction: Greenland and the Question of Sovereignty," *New Global Studies* 7, no. 2 (2013): 117–128.

101. See Kristian H. Nielsen, "Transforming Greenland: Imperial Formations in the Cold War," *New Global Studies* 7, no. 2 (2013): 129–154.
102. Norwegian Ministry of Foreign Affairs. *Regjeringens nordområdestrategi* [The Government's High North Strategy] (Oslo: Norwegian Ministry of Foreign Affairs, 2006).
103. Jonas Gahr Støre, "Iceland and Norway—Neighbours in the High North."
104. Interviews with U.S. defense officials, September 23, 2009.
105. See "The Ilulissat Declaration."
106. See Klaus Dodds and Valur Ingimundarson, "Territorial nationalism and Arctic geopolitics."
107. Nikolaj Petersen, "The Arctic as a New Arena for Danish Foreign Policy: The Ilulissat Initiative and Its Implications," in *Danish Foreign Policy Yearbook 2009*, ed. Nanna Hvidt and Hans Mouritzen (Copenhagen: DIIS—Danish Institute for International Studies, 2009), 57.
108. See Valur Ingimundarson, "Frá óvissu til upplausnar. 'Öryggissamfélag' Íslands og Bandaríkjanna 1991–2006," 39–40.
109. On the U.S. military withdrawal from Iceland, see Ingimundarson, "Frá óvissu til upplausnar. 'Öryggissamfélag' Íslands og Bandaríkjanna 1991–2006," 1–66; Valur Ingimundarson, "Confronting Strategic Irrelevance: The End of a US-Icelandic Security Community?" 66–71; Gunnar Þór Bjarnason, *Óvænt áfall eða fyrirsjáanleg tímamót? Brottför Bandaríkjahers frá Íslandi. Aðdragandi og viðbrögð* [An Unexpected Shock or a Predictable Turning Point? The Withdrawal of U.S. Troops from Iceland: Prehistory and Reaction] (Reykjavík: University of Iceland Press, 2008).
110. "Mikilvægasta verkefnið" [The Most Important Task], *Þjóðviljinn*, December 30, 1956; "Herstöðvaandstæðingar fagna brottför hersins" [Base Opponents Celebrate the Withdrawal of the [U.S.] Military], *Vísir*, October 1, 2006, accessed February 2, 2024, https://www.visir.is/g/20061839069d; "Fagna sögulegum tímamótum með brottför hersins" [Celebrate the Military Withdrawal as a Historical Moment], *Morgunblaðið*, September 29, 2006, accessed February 2, 2024, https://ww.mbl.is/greinasafn/grein/1105325/?t=926461552&_t=1706893262.5667615.
111. Interview with a high-level Icelandic foreign policy official, January 29, 2008.
112. Valur Ingimundarson, "Frá óvissu til upplausnar. 'Öryggissamfélag' Íslands og Bandaríkjanna 1991–2006"; interview with a high-level Icelandic foreign policy official, January 29, 2008.
113. See *Áhættumatsskýrsla fyrir Ísland. Hnattrænir, samfélagslegir og hernaðarlegir þættir* [A Risk Assessment Report on Iceland: Global, Societal, and Military Factors] (Reykjavík: Icelandic Ministry for Foreign Affairs, 2009).

114. Icelandic Ministry for Foreign Affairs, "Samkomulag um samstarf á sviði öryggismála, varnarmála og viðbúnaðar milli Noregs og Íslands" [Memorandum of Understanding between Norway and Iceland on Security, Defense and Preparedness], April 2007, accessed May 25, 2023, https://www.utanrikisraduneyti.is/media/Frettatilkynning/MOU_-_undirritun.pdf.
115. Icelandic Ministry for Foreign Affairs, "Yfirlýsing lýðveldisins Íslands og konungsríkisins Danmerkur um samstarf í víðari skilningi um öryggis- og varnarmál og almannavarnir" [Joint Declaration between the Republic of Iceland and the Kingdom of Denmark on Broader Cooperation on Security and Defense and Civil Preparedness], April 2007, accessed May 25, 2023, https://www.utanrikisraduneyti.is/media/Frettatilkynning/Yfirlysing_Islands_og_Danmerkur.pdf.
116. Icelandic Ministry for Foreign Affairs, "Samkomulag um samstarf á sviði varnar- og öryggismála milli breska konungsríkisins og Íslands" [Agreement between the United Kingdom and Iceland on Defense and Security Cooperation], May 2008, accessed May 25, 2023, https://www.utanrikisraduneyti.is/media/PDF/UK-Iceland_MoU_-Icelandic.pdf.
117. Icelandic Ministry for Foreign Affairs, "Samkomulag milli utanríkisráðuneytis Íslands og varnarmálaráðuneytis Kanada um samstarf í varnarmálum" [Agreement between the Icelandic Foreign Ministry and the Canadian Defence Ministry on Defense Cooperation], October 14, 2010, accessed May 25, 2023, https://www.stjornarradid.is/media/utanrikisraduneyti-media/media/Frettatilkynning/Ice-Can-MOU-final-Icelandic.PDF.
118. See "Stefnuyfirlýsing ríkisstjórnar 2007 [Sjálfstæðisflokks og Samfylkingar]" [The Government Platform of the Independence Party and the Social Democratic Alliance, 2007], May 23, 2007, accessed January 5, 2022, https://www.stjornarradid.is/rikisstjorn/sogulegt-efni/um-rikisstjorn/2007/05/23/Stefnuyfirlysing-rikisstjornar-2007/.
119. See *Skýrsla Ingibjargar Sólrúnar Gísladóttur utanríkisráðherra um utanríkis- og alþjóðamál* [Report by Foreign Minister Ingibjörg Sólrún Gísladóttir on Foreign and International Affairs], *Þingtíðindi*, April 2008, accessed January 5, 2022, 47, https://www.althingi.is/altext/135/s/0857.html.
120. "Geir lýsti yfir óánægju með rússaflug" [PM Geir [Haarde] Expressed Displeasure of Russian Flights], *Vísir*, accessed June 23, 2023, https://www.visir.is/g/200880405055/geir-lysti-yfir-oanaegju-med-russaflug; see also *Áhættumatsskýrsla fyrir Ísland. Hnattrænir, samfélagslegir og hernaðarlegir þættir*.

121. See Jaanus Piirsalu, "Russian warplanes cannot switch on transponders," *Postimees*, September 6, 2016, accessed May 14, 2023, https://news.postimees.ee/3826371/russian-warplanes-cannot-switch-on-transponders.
122. "Opening Address at the Symposium of the Law of the Sea Institute of Iceland on the Legal Status of the Arctic Ocean, delivered by Foreign Minister Ingibjörg Sólrún Gísladóttir," Reykjavík, November 9, 2007, accessed May 20, 2023, https://www.mfa.is/news-and-publications/nr/3983.
123. The author chaired the Risk Assessment Commission.
124. See *Áhættumatsskýrsla fyrir Ísland. Hnattrænir, samfélagslegir og hernaðarlegir þættir.*
125. Icelandic Ministry for Foreign Affairs, "NATO summit welcomes Iceland's initiative on the High North."
126. See *Regjeringens nordområdestrategi.*
127. This decision was made public at a NATO seminar on the High North in Reykjavík in January 2009. On the seminar, see Sven G. Holtsmark and Brooke A. Smith-Windsor, eds., *Security Prospects in the High North: Geostrategic Thaw or Freeze?* (Rome: NATO Defense College, 2009).
128. Interviews with Western officials, December 15, 2009.
129. See Sven G. Holtsmark and Brooke A. Smith-Windsor, *Security prospects in the High North.*
130. See Karl Deutsch, *Political Community and the North Atlantic Area* (Princeton: Princeton University Press, 1957); see also Emanuel Adler and Michael Barnett, "A framework for the study of security communities," in *Security Communities*, ed. Adler and Barnett (Cambridge: Cambridge University Press, 1998), 29–65.
131. See Karl Deutsch, *Political Community and the North Atlantic Area* (Princeton: Princeton University Press, 1957); Emanuel Adler and Michael Barnett, "A framework for the study of security communities," in *Security Communities*, ed. Emanuel Adler and Michael Barnett (Cambridge: Cambridge University Press, 1998), 29–65.
132. Interviews with Permanent Representatives to NATO and Senior NATO officials, December 10 and 12, 2012.
133. Mette Eilstrup-Sangiovanni, "Uneven Power and the Pursuit of Peace: How Regional Power Transitions Motive Integration," *Comparative European Politics* 6, no. 1 (2008): 102.
134. Thorvald Stoltenberg, *Nordic cooperation on foreign and security policy. Proposals presented to the extraordinary meeting of Nordic foreign ministers in Oslo*, February 9, 2009, accessed June 15, 2023, https://www.regjeringen.no/globalassets/upload/ud/vedlegg/nordicreport.pdf; see also Kristin Haugvik and Ulf Sverdrup (eds.), "Ten Years On: Reassessing the

Stoltenberg Report on Nordic Cooperation" (NUPI, IIA, FIIA, DIIS, UI: Oslo, Helsinki, Copenhagen, Stockholm, and Reykjavík, 2019); Clive Archer, "The Stoltenberg Report and Nordic security: big idea, small steps," *Danish Foreign Policy Yearbook* (Copenhagen: DIIS, 2010), 43–74.

135. See Klaus Dodds and Valur Ingimundarson, "Territorial nationalism and Arctic geopolitics"; Ingimundarson, "Territorial Discourses and Identity Politics: Iceland's Role in the Arctic," in *Arctic Security in an Age of Climate Change*; see also Alyson J.K. Bailes and Lassi Heininen, *Strategy Papers on the Arctic or the High North: A comparative study and analysis.*
136. Thorvald Stoltenberg, *Nordic cooperation on foreign and security policy. Proposals presented to the extraordinary meeting of Nordic foreign ministers in Oslo.*
137. See "Stefnuyfirlýsing ríkisstjórnar [Framsóknarflokks og Sjálfstæðisflokks]" [The Government Platform of the Progressive Party and the Independence Party], May 22, 2013, accessed November 5, 2023. https://www.stjornarradid.is/media/stjornarrad-media/media/Rikjandi_rikisstjorn/stefnuyfirlysing-23-3-2013.pdf.
138. See, for example, the comments by Martin Wolf in the *Financial Times*, January 15, 2010.
139. See, for example, Ögmundur Jónasson, "Norræn þögn sama og norrænt samþykki" [Nordic Silence Equals Nordic Approval], November 18, 2009, accessed February 4, 2024, https://www.ogmundur.is/is/greinar/norraen-thogn-sama-og-norraent-samthykki.
140. See *Skýrsla Gunnars Braga Sveinssonar utanríkisráðherra um utanríkis- og alþjóðamál* [Report by Foreign Minister Gunnar Bragi Sveinsson on Foreign and International Affairs]. *Þingtíðindi*, March 2015. https://www.althingi.is/altext/144/s/1074.html; see also "Heræfing NATO á Íslandi," *RUV*, February 1, 2014, accessed July 13, 2023, https://www.ruv.is/frettir/innlent/heraefing-nato-a-islandi.
141. See Björn Bjarnason, *Nordic Foreign and Security Policy 2020: Climate Change, Hybrid & Cyber Threat and Challenges to the Multilateral, Rules-Based World Order*, July 2020, accessed November 14, 2023, https://www.regjeringen.no/globalassets/departementene/ud/vedlegg/europapolitikk/norden/nordicreport_2020.pdf.
142. *Nordic Foreign and Security Policy 2020: Climate Change, Hybrid & Cyber Threat and Challenges to the Multilateral, Rules-Based World.*
143. United States Coast Guard, *Arctic Strategic Outlook* (Washington, D.C.: U.S: Coast Guard Headquarters, 2019), accessed July 20 2023, https://www.uscg.mil/Portals/0/Images/arctic/Arctic_Strategic_Outlook_APR_2019.pdf.; Mike Pompeo, "Looking North: Sharpening America's Arctic Focus," U.S. Department of State, May 6, 2019, https://2017-2021.state.gov/looking-north-sharpening-americas-arctic-focus/.

144. *Nordic Foreign and Security Policy 2020: Climate Change, Hybrid & Cyber Threat and Challenges to the Multilateral, Rules-Based World.*
145. *Nordic Foreign and Security Policy 2020: Climate Change, Hybrid & Cyber Threat and Challenges to the Multilateral, Rules-Based World.*
146. Nordic Council, "Nordic Council calls for closer co-operation on foreign and security policy," Nordic Council, February 9, 2021, accessed 14 February2023,https://www.norden.org/en/news/nordic-council-calls-closer-co-operation-foreign-and-security-policy.
147. See, for example, Gareth Jennings, "Nordic countries combine combat air power," *Janes*, March 24, 2023, accessed June 3, 2023, https://www.janes.com/defence-news/news-detail/nordic-countries-combine-combat-air-power.

CHAPTER 3

Iceland's Financial Crisis and Arctic Identity Politics

The October 2008 banking collapse in Iceland—at the beginning of the global financial crisis—had profound effects on its relations with the outside world. Under intense domestic political pressure, the coalition government of the Independence Party and the Social Democratic Alliance quickly began to prioritize economic and societal security at the expense of traditional military security. Initially, there was limited understanding for Iceland's economic plight among its U.S., European, and Nordic allies who—with more than a hint of epicaricacy—saw the crash largely as a self-made disaster. They pointed out that through toothless financial supervision, Icelandic political elites had refused to listen to repeated foreign warnings and continued to promote an unsustainable banking expansion abroad.[1] When Iceland's main Western partners turned down its requests for emergency aid, the Icelandic government felt compelled to turn, formally, to Russia, asking it for a $5.4 billion loan. The conservative Prime Minister Geir Haarde put it bluntly: "We have not received the kind of support that we were requesting from our friends," and in such a situation, "one has to look for new friends."[2] This abrupt call for new loyalties was not only fueled by political despondency; it was also influenced by hedging considerations aimed at opening up alternative non-Western venues in Iceland's foreign policy to respond to a state of exception without abandoning its basic Western orientation. It was, however, difficult to accomplish such a goal in the absence of deeper political traditions and institutional ties.

© The Author(s), under exclusive license to Springer Nature Switzerland AG 2024

V. Ingimundarson, *Iceland's Arctic Policies and Shifting Geopolitics*,
https://doi.org/10.1007/978-3-031-40761-1_3

This is not to say that such a move lacked historical precedents. The Soviet Union had briefly become one of the main trading partners immediately following World War II and again—despite Iceland's Western political and military integration during the early Cold War—in the first half of the 1950s after it filled a vacuum created by a British landing ban on Icelandic fish, which was imposed in retaliation for Iceland's unilateral extension of its fishery limits. Indeed, Iceland's barter trade with the East bloc became 35% of its total trade in the latter half of the decade—a development that put it in a special category, with the corresponding figure for most other Western countries being only 3–4%.[3] Even if the liberalization of the Icelandic economy in the 1960s—with the abolition of import and export controls—led to a sharp reduction in the barter trade, the Soviet Union retained a 10% share of Iceland's total trade until the end of the Cold War. The Soviets bought Icelandic fish in exchange for oil, timber, and other raw materials. Iceland's solidarity with Western policies—which was severely tested during the three "Cod Wars" with the United Kingdom from the 1950s through the mid-1970s—was, in fact, always conditioned by a willingness to protect its fish-for-oil deal with Russia. After the disintegration of the Soviet Union, however, Iceland's bilateral trade relations with Russia became negligible.[4]

In October 2008, the Russians reacted, at first, positively to Iceland's loan entreaty, which had first been broached by Icelandic government officials during the summer due to a rapidly deteriorating financial situation or before the banking crash. Prime Minister Vladimir Putin allegedly personally approved a Russian loan to Iceland.[5] But Russian government ministries, which had mostly been kept in the dark about the matter, dragged their feet, even if Finance Minister Alexei Kudrin claimed, publicly, that there was a willingness to assist Iceland.[6] One reason for the reluctance to make a firm commitment was the decision by the head of the Icelandic Central Bank, former Prime Minister Davíð Oddsson, to make the prospective loan agreement public without informing the Russians beforehand. Even if he quickly qualified his statement by saying that negotiations were still ongoing,[7] the episode raised unfounded suspicions on the part of the Russians about an Icelandic double game designed to attract Western economic assistance to forestall a Russian loan.[8] When the Icelandic government concluded that it could not wait any longer for a Russian reply, it turned to the IMF for an emergency rescue loan. The move was strongly opposed by Oddsson, who believed that the IMF's controversial record—with its strict conditionality policies attached to loans—would infringe on Iceland's sovereignty. The politician-turned-banker Oddsson was a divisive figure after a long political career. He continued to wield influence within the Independence Party, which he had dominated during his 13-year tenure as Prime Minister, but he was highly unpopular among its coalition partner, the Social Democratic Alliance, which also had deep misgivings about the overture to Russia. After the Icelandic government approached the IMF, the Russian government saw no reason to finance a rescue package.[9]

While the Russian gambit did not lead to anything, it raised larger questions about the instability of Iceland's foreign policy, Cold War legacies, and Arctic futures. Iceland faced criticisms for its eagerness to accept aid from Russia—and in the Western media, the loan was immediately framed in anti-Western terms. It was argued that Russia wanted to use Iceland as part of a strategy to bolster its claim to Arctic oil and gas resources.[10] That Russia was keen on expanding its geostrategic influence in the North was plausible, but Iceland's Arctic role was limited and had nothing to do with Russian jurisdictional claims in the region. The debate also led to Western media speculations about Iceland's intention of offering Russia base rights as a sign of gratitude for a potential loan. What gave them added currency was a political intervention by Iceland's President, Ólafur Ragnar Grímsson, who at a Danish embassy function—where the Russian, American, and British ambassadors were, among others, present—sharply criticized Iceland's traditional allies, especially Britain, for using anti-terrorist legislation to freeze the assets of the failed Icelandic bank subsidiary in the United Kingdom but also the United States and the Nordic countries for their refusal to offer Iceland financial assistance. Echoing Haarde, Grímsson stated that Iceland needed new friends, but he went much further by suggesting that Russia be invited to take over the former U.S. base in Iceland.[11] Taken aback, the Russian Ambassador, Viktor Tatarintsev, replied that "at this point," the Russians did not need a base for military aircraft in Iceland. Grímsson's comments attracted considerable public attention after a memorandum by Norway's ambassador to Iceland on the discussion was leaked to the media.[12] In it, there was, however, no mention of Tatarintsev's qualification, which explains why his reaction failed to receive attention.[13]

Having no formal foreign policy competency, President Grímsson was, it turned out, acting on his own by venting his moral indignation over lack of Western support for Iceland in an emergency situation. There was no Icelandic government plan to offer the Russians base rights which would—without doubt—have evoked domestic political opposition and be interpreted by Iceland's Western partners as an affront to its NATO membership. To float such an idea in a diplomatic setting was reckless, coming on top of the previous risk-laden initiative Icelandic ministers and officials took when they sought a loan from Russia, which amounted to one-third of Iceland's GDP. Historically, it would not have been the first time that Iceland—which topped the list of U.S. per capita foreign assistance from the late 1940s until the mid-1960s[14]—would use its strategic location to play off East against the West. In the mid-1950s, NATO provided Iceland—in a first for a member country—with economic aid to prevent it from accepting a huge loan offer from the Soviet Union.[15] But in 2008, Iceland was in no bargaining position to court both sides, simultaneously.

The Icelandic government's aim was not to prepare the ground for a foreign policy shift or to abandon the country's traditional political, economic, and cultural ties with the West by joining a Russian alliance.[16] But what these hastily assembled hedging initiatives betrayed was the state of mind of a panic-stricken and fearful political elite, scrambling for ways to prevent an economic collapse. The overtures to Russia had an indirect effect on Iceland's security policy discourse, with the Icelandic government ceasing any public criticism of Russia's bomber flights near the country. Up to then, the Russians had refrained from responding to it, viewing it more as an irritant than a major problem in the bilateral relationship, but they had planned a formal diplomatic rebuttal, if the Icelanders stepped up their recriminations.[17] After the financial crash, there was no need to do so since the Icelandic government was immersed in an economic salvage operation, which overshadowed any hard security considerations or Cold War residues about Russia as a military threat.

3.1 Embellishing the Economic Promise of the Arctic

The financial crisis led to the downfall of the Conservative-Social Democratic coalition government in January 2009. It was replaced by a minority Left-Wing coalition government—composed of Social Democrats and the Left Greens with the parliamentary backing of the Progressive Party—which remained in power after winning a landslide victory in parliamentary elections in April. Among its key foreign policy goals were to apply for EU membership and to raise Iceland's Arctic profile. The prioritization of both issues must be put within the context of the response to the financial crisis. Initially, the most logical outcome of the "politics of transition" was seen by many as a shift toward the European Union. The EU application signified a belated effort to restore economic stability at home and political backing among Western states abroad. It was, however, deeply controversial. Pushed by the Social Democrats as a precondition for government participation, their coalition partner, the Left Greens, reserved the right to reject Iceland's EU accession in a referendum.[18] In addition, the conservative and centrist opposition parties, the Independence Party and the Progressive Party, turned against Iceland joining the EU.[19]

Iceland's admission to the European Union mattered, of course, far less to the EU, but from the perspective of some member states, it could have a stabilizing effect on the Icelandic economy and facilitate the solution of the diplomatic dispute, emanating from the banking crash, with Britain and the Netherlands. It was about whether the Icelandic government was liable or not for the so-called Icesave accounts, individual savings accounts—offered by the Icelandic bank subsidiary of the collapsed Landsbanki—in these countries. The row complicated Iceland's efforts to negotiate a rescue package with the IMF and to repair relations with Western countries.[20] From a geopolitical perspective, the argument in favor of Iceland's EU membership was that it could add to the Union's strategic reach through enhanced presence in the Arctic. Yet, unlike the Arctic turn in Iceland's foreign policy, the EU application exposed divisions over its foreign policy orientation. While the EU accession process was on track during the tenure of the Left-Wing government or until the spring of 2013, it was abruptly abandoned by the Center-Right coalition government of the Independence Party and the Progressive Party, which succeeded it.

From 2009 on, the Arctic took on a new functional and psychological dimension as part of the Left-Wing government's handling of the financial crisis. Compared to some other Arctic states, Iceland was, surely, not among those that were poised to gain much from natural resource extraction. Still, the Icelandic government prepared for oil exploration in the Dragon Area—in its EEZ near Jan Mayen, an island under Norwegian sovereign control—through a licensing process for foreign companies. The initiative dated back to the 1980s, when Iceland and Norway began joint scientific research in the Jan Mayen area to explore its oil potential;[21] in 1999, the Icelandic government had formed an interagency committee to study the matter further, and in 2005, it was decided that formal procedures for natural resource extraction in this area should be initiated.[22]

Subsequently, the idea was linked to potential economic opportunities after the financial crash, such as making Iceland a service center in connection with oil and gas exploitation and tourism in Greenland and a future transshipment hub. Thus, the prospects of economic gains in the Arctic and the emergence of new sea routes could provide an antidote to the traumatic effects of the banking collapse.

3.2 Iceland's First Arctic Policy: Working Toward a Non-militarized Region

The Social Democratic Foreign Minister Össur Skarphéðinsson was responsible for cementing the Arctic region as a core component of Iceland's foreign policy during his term in office from 2009 to 2013. Shortly after he became minister, the Icelandic Foreign Ministry published a report on the Arctic, underpinning Iceland's stakeholding claim and insisting on a "seat at the table" where decisions were made. In addition, it stressed that what underpinned Arctic governance was peaceful cooperation of Arctic states, with the Arctic Council playing a key role. Hence, there was no need to create new institutions or to conclude international treaties on the Arctic. This was an obvious dig at the much-maligned Arctic Five initiative as well as the European Parliament's idea of an Arctic Treaty based on the Antarctic precedent. To inflate Iceland's Arctic position and to put forward a "status-marker," Skarphéðinsson even claimed—in the report's preface—that Iceland was the only country being located in its entirety within what was "generally" defined as the Arctic Circle. Even if there are many definitions of the Arctic region, the one based on the Arctic Circle accorded mainland Iceland no such status[23] because it only included Grímsey Island. Apart from portraying Iceland as an ideal location for a transshipment hub, the report touted it as a potential service center for natural resources extraction and Arctic shipping.[24] Thus, the transarctic polar sea route between the North Atlantic and the Pacific through the Atlantic Ocean was offered as a realistic possibility due to reduced ice and the development of new shipping technologies. Since the focus at that time was on the Northern Sea Route along Russia's northern coastline, it was a gross overstatement. Still, it served the political aim of elevating Iceland's location in connection with alternative future Arctic transport routes.

As a sign of Iceland's preoccupation with the Arctic, the government began, in 2010, to pay special attention to the political, economic, and legal dimensions of Arctic governance and to diplomatic overtures by non-Arctic Asian states, notably China, and regional bodies, such as the European Commission, to become observers in the Arctic Council. To retain a decision-making role in Arctic affairs, Iceland continued to resist the development of a hegemony exercised by the Arctic Five. Thus, it protested strongly when Canada decided to follow up on the Ilulissat meeting by convening another one in Chelsea, Canada, in 2010. Skarphéðinsson rejected an offer—extended by the Canadian hosts to Iceland, Finland, and Sweden—to have an observer status at the meeting.[25] He received, it turned out, an unexpected boost from U.S. Secretary of State, Hilary Clinton, when she used her attendance at the Chelsea meeting to reprimand, publicly, the Canadian government for not inviting other Arctic stakeholders, that is, the representatives of the Indigenous peoples and the three non-littoral Arctic states.[26] Even if no further Arctic Five Foreign Ministers meetings took place after the one in Chelsea, officials from the five states continued to meet to discuss Arctic issues in the following years.[27]

A comprehensive Arctic policy—developed by Skarphéðinsson with cross-political backing—was completed in the form of a parliamentary resolution, which was approved in early 2011.[28] In general, the policy straddled a line between the Arctic as an ecological frontier to be regulated by an international regime, on the one hand, and a high-stakes resource base and a geopolitical arena, on the other. It promised to fight human-made climate change to improve the lives of those who lived in the Arctic and their communities. Since other Arctic states had already published their strategies, it was only natural that Iceland would want to do the same.[29] The timing was, however, not specifically related to such initiatives.[30] The Icelandic Foreign Ministry had published reports in 2005 and 2009, which reflected Iceland's Arctic positions.[31] Yet Skarphéðinsson's idea was far more ambitious. He wanted to receive a mandate to move the Arctic to the center of Iceland's foreign policy, which he succeeded in doing, when the parliamentary resolution was passed.

The Arctic policy was based on several core points.[32] While Iceland's approach toward the North was driven by its strategic location, it was also geared toward multilateralism and identity politics in relation to other Arctic states. Given Iceland's membership in the UN and the Arctic Council as "status-markers," the policy made a firm commitment to

UNCLOS and to the AC as the legitimate institutional governing forum for the area.[33] It was an effort to protect Iceland's identity as a core Arctic state. The argument was made that the AC should be provided with more political weight and normative regulatory instruments to deal with Arctic shipping, natural resource extraction, and tourism. The idea was to elevate the Arctic Council from a pure decision-shaping intergovernmental body to a decision-making one with respect to issues such as search and rescue and anti-pollution measures. It fitted with the portrayal of the eight Arctic Council states—together with Arctic Indigenous peoples—as equal stakeholders, enjoying a privileged position in the region, even if it did not entail opposition to the involvement of non-Arctic states and organizations in the Council's work.[34]

The policy also emphasized the need to expand ties with Greenland and the Faroe Islands as part of West Nordic cooperation in the fields of trade, energy, resource extraction, environment, and tourism. Ever since the West Nordic Council—as a venue for parliamentary cooperation between Iceland, the Faroe Islands, and Greenland—had been established in 1985, successive governments had cultivated it. The forum was seen as part of Iceland's wider marine resource interests in the North Atlantic and political and cultural relations with its closest neighbors.[35] To strengthen such tripartite cooperation in the Arctic would, according to the policy, help elevate their West Nordic common international, economic, and security-political status. The goal was to create a balance between exploitation and conservation: to harness economic opportunities, while promising, simultaneously, to exercise environmental responsibility. Thus, it was pointed out that the future extraction of Greenland's oil deposits and valuable minerals required a strong regulatory regime to protect the marine environment. Yet, with a reference to Icelandic construction companies operating in Greenland, the benefits of increased bilateral political and trade relations, including in the field of civil aviation, were the underlying theme. Similarly, the 2005 Hoyvík Free Trade Agreement on the creation of a common market with the Faroe Islands—which was the most extensive external trade agreement negotiated by Iceland with a foreign political entity—was singled out for praise.[36] While the Faroe Islands were an important partner in West Nordic cooperation, Iceland's main focus was on Greenland within a specifically Arctic context.[37]

In addition, the policy expressed strong support for the rights of the Indigenous peoples in the Arctic. They should have a say in all matters pertaining to their political, social, ecological, economic, and cultural interests and be guaranteed the right to protect their cultural uniqueness, to strengthen their social structures, and to improve their living conditions.[38] There was no engagement with specific Indigenous groups—in the policy—or references made to Indigeneity when discussing Greenland and its economic potentials. It sounded, furthermore, self-serving to criticize the Arctic Five for snubbing Indigenous peoples because the Icelanders were mainly thinking about their own exclusion from the meetings. Still, despite these flaws, the text was emphatic in its backing for Indigenous rights.

Having been under foreign rule for centuries, Iceland had—as a small state and a young republic—generally supported the right of self-determination of peoples with sovereign aspirations. To be sure, from a historical perspective, Iceland had not been perfectly consistent on this issue when it came to Greenland. Having written a doctoral thesis on the legal status of Greenland in the Middle Ages, Icelandic economist, Jón Dúason, argued vocally—in the interwar period and after World War II—that Iceland should make an ownership claim to it. A committee of legal experts—appointed by the Icelandic Foreign Minister Bjarni Benediktsson—concluded that Iceland had no such right. Iceland was, thus, not prepared to challenge the Danish legal position on Greenland. When Greenland was incorporated into the Kingdom of Denmark in 1953 and became a non-colonial subject within the Danish Realm, Iceland abstained in the UN vote on the island's new status in 1954.[39] In the 1990s, however, Iceland was the first state to recognize the restored independence of the Baltic states before the disintegration of the Soviet Union. It was also one of the first to recognize the new states that came into being following the break-up of Yugoslavia with the exception of Kosovo.[40] When it came to the Arctic, the position on self-determination was driven less by principles than by specific political and economic interests. On the one hand, it was thought that any hegemonic aspirations on the part of the Arctic Five would be undercut by excluding the Arctic Indigenous people; on the other hand, Iceland wanted to support Greenland's status-seeking efforts, including an independence drive, while, at the same time, benefit economically from its geographic proximity.[41]

The Icelandic government followed through on its aim to strengthen the relationship with Greenland. Before the approval of Iceland's Arctic policy, the two sides had already in 2008 agreed to set up a committee to discuss joint projects. After the conclusion, in 2009, of the Act on Greenland Self-Government—which expanded on the 1979 Home Rule Act by transferring further competencies and responsibilities from the

Danish to the Greenlandic authorities—the political circumstances for forging stronger bilateral ties were advantageous. It was, however, not until Skarphéðinsson took the initiative by developing Iceland's Arctic policy that things began to change. In addition to agreements on health and fishery cooperation, an Icelandic-Greenlandic chamber of commerce was created. Most importantly, Iceland established a consulate in Greenland's capital, Nuuk, in 2013, with the appointment of a Consul General with ambassadorial rank. That the Icelanders were thinking about material benefits from this relationship was clear from the start. Skarphéðinsson saw the eastern part of Iceland as a key area for servicing future oil extraction in Eastern Greenland and the tourism industry as a promising area of cooperation.[42]

Iceland's Arctic policy defined security in broad terms and tied it to civil threat scenarios, such as accidents at sea, increased maritime traffic, and pollution caused by exploitation of natural resources. In an explanatory note, it was acknowledged that incremental securitization had taken place in the Arctic as part of "sovereignty controls" and monitoring. To work against the militarization of the Arctic, transnational cooperation in fields such as search and rescue was encouraged. Support was also expressed for increased Nordic security cooperation in this area, with reference to the Stoltenberg report. And as a model for security agreements, the policy referred to arrangements Iceland had concluded with Norway, Denmark, Britain, and Canada following the closure of the U.S. military base.[43] These agreements, however, were, as noted, modest, since they did not exceed any NATO obligations, had no specific bearing on the U.S.-Icelandic Defense Agreement, and entailed no security guarantees of any sort.

Reflecting a status-seeking attempt to strengthen Iceland's position in the Arctic Council and to undercut the Arctic Five stewardship claim in the Arctic Ocean, the policy also made the case for counting Iceland among the Arctic littoral states. It rejected a narrow legal reading of the term "coastal state," a definition based on the delimitation of the Arctic continental shelf according to UNCLOS and on being restricted to the Arctic Ocean itself. It was argued that since its EEZ extended to the Arctic Greenland Sea—which has been defined as either being part of the Arctic Ocean or bordering it—Iceland could make a case for a coastal state status.[44] In addition, the purpose was to define Iceland's place in the Arctic in broad political and geographic terms. It was not only meant to remind both international and domestic audiences of the essence of Arctic geopolitics—the inside/outside divide, with Arctic states pitted against non-Arctic states. Such a speech act also fitted well with ideas about looking at the borders of Arctic states as moving in two areas simultaneously: to sites inside sovereign territories and offshore or within fixed spaces as well as provisional ones. UNCLOS had provided the opening for such a reading by its formula for sovereign state rights and jurisdiction beyond the 200-mile EEZ regime.

The nationalist assertiveness contained in the Arctic policy was elite-driven.[45] Politicians were not responding to domestic political debates or nationalistic passions over the Arctic.[46] The projection of Iceland as a coastal state was about cementing and "naturalizing" its Arctic identity. As has been touched up on, when Icelandic officials began to popularize the Arctic and its potential in the mid-2000s, they resorted to celebratory frontier narratives in which historical accounts of navigation were mixed with future possibilities offered by the shortening of transport routes between Asia, Europe, and North America.[47] Yet metaphors about modern-day Viking territorial conquerors had to be unceremoniously abandoned after the financial crash since they had figured prominently in the promotion of Iceland's disastrous international banking expansion. The reputational damage caused by the crisis was hard to accept by Icelandic elites.[48] It destabilized and undercut preponderant identities and self-perceptions about Iceland as a developed, affluent Western state.[49] Hence, the importance of defending Iceland's position in the Arctic as part of efforts to reclaim its standing abroad.

To engage in this status-seeking exercise—and to enhance Iceland's influence within the Arctic's preeminent forum—Skarphéðinsson decided, in 2011, to take on Norway over the hosting of the new Arctic Council secretariat, whose establishment 15 years after it had been initially proposed was agreed on at an AC ministerial meeting in Nuuk the same year. Prodded by the Americans, the Icelanders offered Reykjavík as the secretariat's home site, stressing its central and accessible location for all Arctic states.[50] The Norwegians, however, pressed for its "Arctic capital"—Tromsø—which was eventually selected despite misgivings by some AC members due its distant Northern location. The Icelanders were informed that Norway's active involvement in NATO's 2011 military intervention in Libya had prompted U.S. Secretary of State Hilary Clinton to switch sides and withdraw U.S. support for the Icelanders in favor of the Norwegians.[51] This was a classic case of what has been termed "bandwagoning for status," with the Norwegians capitalizing on social recognition accorded to them by a Great Power.[52]

Partly to placate the Icelanders, the Norwegians agreed to expand academic cooperation with Iceland by funding a new professorship at the University of Akureyri, named after the Norwegian explorer, Fridtjof Nansen, through an agreement between the Icelandic and Norwegian Foreign Ministries in 2011.[53] As the second largest urban center outside the Reykjavík area, Akureyri had marketed itself, like Tromsø in Norway, as an "Arctic capital"—where, in addition to the university, the Stefansson Arctic Institute as well as two Arctic Council Working Group secretariats, PAME and CAFF, were located. Despite losing out to Norway over the location of the Arctic Council secretariat, Iceland benefitted, in the end, from its establishment: Magnús Jóhannesson, the former State Secretary of Iceland's Environmental Ministry, was appointed the first Director of the Arctic Council secretariat when it was set up in Tromsø in 2013.[54] What proved instrumental in favor of his candidacy was that the larger Arctic states, notably the United States, could not agree on Russia's candidate Anton Vasiliev, its Senior Arctic Official,[55] whom the Norwegians backed in exchange for Russia's support for its bid for Tromsø.[56] Vasiliev, incidentally, became, in the following year, Russian Ambassador in Iceland, where he served for six and a half years.[57]

In addition to plans for oil exploration in its EEZ near Jan Mayen, the Icelandic government continued to offer Iceland as a transshipment hub in connection with the future opening of a transarctic route. It also stressed increased West Nordic cooperation in a move that was not only meant to cultivate friendly, political, cultural, and economic relations with Greenland and the Faroe Islands. It was directly tied to natural resource politics in Greenland—to provide services to the "energy triangle."[58] At the same time, the government wanted to complement its material aspirations by addressing the potential effects of climate change on the marine environment in the North through more investments in maritime surveillance and resource evaluation. It was an admission that oil and gas shipments could lead to environmental risks, especially oil spills. Increased involvement of the International Maritime Organization (IMO) in the Arctic was seen by the Icelanders as a step in the right direction. The perceived need to strengthen international and regional regulatory frameworks reflected Iceland's vulnerability to sudden changes in the marine eco-system, stemming from climate change or ecological pollution. Hence, Iceland fully supported the Search and Rescue (SAR) and anti-pollution intergovernmental agreements, which were negotiated in 2011 and 2013, respectively, between the Arctic states under the rubric of the Arctic Council, as well as the 2014 adoption of a legally binding Polar Code through the IMO on the design, construction, and operation of commercial ships in the Arctic.[59] Other IMO conventions, such as the International Convention for the Safety of Lives at Sea (SOLAS) and the International Convention for the Prevention of Pollution from Ships (MARPOL) also covered the Arctic. The Polar Code complements these binding treaties.

While reiterating Iceland's commitment to NATO and the U.S.-Icelandic Defense Agreement, Foreign Minister Skarphéðinsson continued to prioritize non-military threat scenarios in the Arctic.[60] It was a view that was widely shared by political elites in other Arctic states at the time. The 2013 Arctic strategy of the U.S. Department of Defense stated, for example, that there were no state-based threats in the region. The Pentagon even specifically warned against inflated rhetoric about the militarization of the Arctic, boundary disputes, and competition for resources because it could create "an 'arms race' mentality that could lead to a breakdown of existing cooperative approaches to shared challenges."[61] In other words, misperceptions about Arctic geopolitics were seen as a risk factor in its own right. In his quest for using Iceland's strategic location in the North—after the departure of U.S. troops—to highlight a soft security agenda, Skarphéðinsson proposed the establishment of an International Rescue and Response Center at the former military base to provide support capability for emergency operations in the Arctic and to offer facilities and opportunities for joint SAR training.[62] Domestic considerations also played a role: Rescue operations, especially north of Iceland, faced serious limitations with no real harbors along the east coast of Greenland and only a few gravel airports. The proposal was both consistent with the emphasis—in Iceland's Arctic policy—on promoting the Arctic as a peaceful region as well as with its backing for SAR and oil spills prevention measures.

Skarphéðinsson's successor as Foreign Minister, Gunnar Bragi Sveinsson (2013–2016) from the Progressive Party,[63] was also committed to the center. Yet, while the Icelandic government had the idea under active internal consideration for the next several years, no action was taken on it.[64] Questions about the interest of other Arctic states in such a facility and about its purpose and functional role—as well as about whether it should be limited to Iceland's Western security partners or include others, such as Russia—were never resolved. And when the Americans began, in 2016, to use its former Keflavík base area as a staging ground for anti-submarine operations in the North Atlantic and the Arctic, it was clear that soft security considerations had been trumped by hard security ones.

3.3 A Foreign Policy Hedging Experiment: Expanding Relations with China

The absence of American interest in Iceland after the base closure created more space for a small state like Iceland to pursue non-traditional foreign policy options. In studies of hedging, insufficient attention has been given to state efforts to deal with non-military threats.[65] In 2008, the financial crisis in Iceland presented by far the most dangerous security risk. Since the end of the Cold War, Iceland had been part of a security environment in Northern Europe where perceived military threats were considered non-existent.[66] In the years after the banking collapse, the overwhelming emphasis of successive Icelandic governments was on economic reconstruction.

Iceland's increased trade and economic relations with China reflected this preponderant materialist agenda. China and Iceland had already in 2005 signed a protocol, which paved the way for negotiations on a Free Trade Agreement (FTA).[67] Iceland put the talks on hold in 2008 after the banking collapse. Still, as a sign of political good will during the financial crisis, the People's Bank of China signed, in 2010, a bilateral currency swap agreement with the Central Bank of Iceland, with a three-year maturity.[68] While the gesture did not have much effect on Iceland's economic revival, its symbolic value was appreciated by the Icelandic government. The same applied to China's support for an IMF bailout for Iceland without strings attached; it flew in the face of the position of several Western states, which had insisted on the prior settlement of the Icesave issue with Great Britain and the Netherlands.[69]

The President of Iceland, Ólafur Ragnar Grímsson, played a significant public role in promoting Iceland's cooperation with China.[70] Soon after he took office, he had set his sights on cultivating bilateral ties. On Grímsson's invitation, the President of China, Jiang Zemin, had visited Iceland in 2002. Three years later, Grímsson visited China—together with a large delegation mostly made up of a businesspeople—and met with its President Hua Jintao. During the visit, four agreements were signed, including one on environmental cooperation and another one on mobile networks,[71] involving a private Icelandic mobile operator and Chinese technology company Huawei, which was later to become a Western target for its ties with the Chinese government.[72] Foreign Minister Skarphéðinsson was also keen on expanding the China relationship, especially when Iceland faced the stonewalling of Western countries due to the Icesave banking dispute. What followed were high-profile and high-level Chinese ministerial visits to Iceland, including the visit of Premier Wen Jiabao in 2012 and a reciprocal visit by Icelandic Prime Minister Jóhanna Sigurðardóttir in 2013. A bilateral framework agreement on Arctic collaboration, including scientific cooperation,[73] which was signed in 2012, led to the establishment a year later of the China-Nordic Arctic Research Center (CNARC),[74] which originally included four Chinese and six Nordic member institutes. In 2013, the two sides also completed the work on the FTA, with Iceland being the first European country to conclude such a treaty with China.[75]

In addition, there was an affinity between Icelandic and Chinese official Arctic discourses on the future transpolar, or Arctic Central Route, running from the Atlantic Ocean to the Pacific Ocean. The feasibility of this future shipping lane had been highlighted in the 2009 report by the Icelandic Foreign Ministry on the Arctic.[76] It was, however, not taken up by the Icelandic government until 2012–2013,[77] when Skarphéðinsson approached the Chinese about exploring the economic prospects of the transpolar route, with the participation of other potential partners such as the EU and the United States through Alaska.[78] If the transpolar route materialized, the argument went, it would serve Chinese economic interests by bypassing the Russian-controlled Northern Sea Route and Icelandic ones by marketing Iceland as the most logical transarctic transport hub.[79] It was, however, a utopian project, which faced huge logistical difficulties and had no chance of being implemented in the prevailing conditions.

For this reason, the Chinese continued to focus on the Northern Sea Route, especially in view of their longstanding political ties and economic cooperation with Russia, which had eased bureaucratic restrictions on the use of the NSR and lowered transit fees to make it more attractive.[80] Political obstacles remained, however. After Russia's assault on Ukraine in 2022, the Chinese COSCO Shipping Corporation, whose fleet is one of the largest in the world, stopped using the NSR, presumably out of fear that its U.S. business could be hurt due to Western economic sanctions against Russia.[81] On the other hand, a new Chinese shipping company, Newnew Shipping Line, with possible links to COSCO and/or the Chinese government, started in the summer of 2023 a container line between China and Russia along the NSR.[82]

Apart from Iceland's strategic location, China's cultivation of the bilateral relationship was undoubtedly partly motivated by broader Arctic aspirations, such as its application for an observer seat in the Arctic Council.[83] It was underscored, in a politically symbolic way, by the transpolar voyage of the first domestically built Chinese icebreaker Snow Dragon (R/V Xue Long) in 2012, which made Iceland its main destination, and which attracted considerable media attention.[84] Another example was the involvement of China's National Offshore Oil Corporation (CNOOC) in oil exploration in Iceland's Dragon Area.[85] It was consistent with China's interest in developing Arctic resources through its state-run enterprises in partnerships with companies from Arctic countries. As was the case with Iceland's overture to Russia after the financial crash, some commentators jumped on the Icelandic-Chinese link, arguing that it served China's global strategic interests.[86] It also included exaggerations, such as the claim that the Chinese embassy in Reykjavík was the "largest" in Iceland.[87] The Chinese had bought a large building to be sure. Yet they had taken advantage of a depressed real estate market following the financial meltdown by buying it at a rock bottom price, and their staff was not outsized compared to other embassies.[88]

While the Free Trade Agreement was generally welcomed by politicians, it was opposed by the Icelandic Federation of Labor on the grounds that the Chinese government did not respect human and trade union rights.[89] It feared that the agreement could lead to lower wages in Iceland because local companies could not compete with Chinese ones on equal terms. Otherwise, Iceland's expanded economic relationship with China did not meet much domestic political resistance, except in one case. In 2012, the Left Green Minister of the Interior, Ögmundur Jónasson, single-handedly thwarted plans by the Chinese real estate tycoon, Huang Nubo—acting through the Beijing Zhongkun Investment Group—to purchase land in northeastern Iceland, comprising 300 square kilometers or 0.3% of the country's total territory, for the purpose of developing a $200 million tourist resort with a hotel, an airport, and a golf course.[90] To many Icelanders, the intention of using one of the most barren parts of the country to develop a project on such a grand scale sounded farfetched, raising suspicions of hidden Chinese political agendas.[91] Local Icelandic business interests, however, supported the idea on the grounds that it represented a sizable foreign investment.[92]

Whether Nubo had a direct relationship with the Chinese government was unclear. Based on his Communist Party background, some critics saw his venture as a Chinese government proxy attempt to establish an outpost in the Arctic region. Whatever the motive, his bid was turned down on legal grounds, even if political factors were decisive. Only EEA nationals have the right to buy land in Iceland; other foreigners need a special permission, which Interior Minister Jónasson refused to grant in the Huang Nubo case because he was suspicious of Chinese intentions and because he was, in principle, against allowing affluent foreigners to purchase land in Iceland. Prime Minister Jóhanna Sigurðardóttir, Minister for Foreign Affairs Skarphéðinsson, and Minister of Finance Steingrímur J. Sigfússon had been far more positive toward the Chinese businessman's bid. But Jónasson managed to outfox them as they tried to postpone the final decision on Nubo's application for completing the deal by seeking the advice of other lawyers than Jónasson had consulted within his own ministry and by proposing that other ministers have a say on the matter. At a governmental meeting—where the Prime Minister and other ministers made their case—Jónasson informed his colleagues that he had already submitted a rejection letter to the Chinese representatives. Afterward, he was severely criticized by his Social Democratic Alliance's coalition partners, including Sigurðardóttir and Skarphéðinsson, who argued that the decision contradicted the government's policy on foreign investments. Yet, defeated, they did not take any further action.[93]

The "Huang Nubo affair" had, it turned out, no detrimental impact on Iceland's political and economic relations with China. Before Jónasson made his decision, the Chinese ambassador in Iceland had claimed that his government had nothing to do with Nubo's venture—a claim that the Minister of the Interior took with a grain of salt.[94] The Sino-Icelandic agreements on Free Trade Agreement and scientific cooperation in the Arctic were signed after the land purchase was rejected. Despite the setback, Nubo claimed, at first, that he was still committed to his business project in Iceland, seeking a long-term lease instead of a purchase. As part of the plan, Icelandic municipal stakeholders were to buy the land and rent it out to the Chinese businessman to circumvent Jónasson's decision. But before any agreement was concluded, the Icelandic partners announced, at the end of 2014, that nothing would come of the project, citing Nubo's lack of interest.[95] In an ironic twist, a British tycoon, Jim Ratcliff, bought the land in 2016 without any plans to develop it. As an EEA resident, he was entitled to do so without an Icelandic ministerial permission, and his purchase attracted minimum media attention.[96] Jónasson criticized the deal publicly, but he was in no position to do anything against it three years after leaving the government.[97]

Iceland's backing for China's observer status in the Arctic Council was consistent with its support for the applications of the other Asian states, that is, Japan, South Korea, India, and Singapore. It reflected a coordinated Nordic position on non-Arctic observer states, which until then included only seven European states: Germany, France, Italy, the Netherlands, the United Kingdom, Poland, and Spain. When it was decided—at the 2013 Arctic Foreign Ministers meeting in Kiruna, Sweden—to accept the Asian states as observers,[98] the outcome could, on the surface, be seen as an affirmation of the approach adopted by the Nordic states. They had made the case that the addition of observers would strengthen the Arctic Council and discourage non-Arctic states and organizations from seeking new institutional venues for pursuing their interests in the region. Despite this inclusive attitude, however, there was never any intention on the part of Iceland to dilute the primary role of the Arctic Eight in the Council or to give external states or organizations a decision-making voice.[99] On the contrary, the bottom line in Iceland's Arctic strategy was to ensure that the Arctic Five forum would not be able to assume a dominant role in the region at the expense of the Arctic Council.

Iceland's relationship with China did not follow a pre-determined path dictated by the latter's strategic interests. It was, after all, the Icelandic government that had approached the Chinese in the first place because it wanted to deepen the bilateral relationship as part of its post-crisis economic reconstruction agenda. The FTA between China and Iceland was far more limited in scope than those concluded between China and New Zealand, Australia, or Switzerland because it did not cover investments.[100] Chinese investments in Iceland have, in fact, been few and, with one exception, not defined as a Foreign Direct Investment (FDI) in accordance with the guidelines established by the Organisation for Economic Co-operation and Development (OECD)[101] as a minimum 10% ownership stake in a foreign-based company. The Geely Group's 2015 investment of $14.5 million in a recycling company—Carbon Recycling International (CRI)—is the only Chinese FDI in Iceland.[102] The parent company of Elkem Norway, which runs an Iceland-based metal plant that produces ferrosilicon, is China National Bluestar, but the venture is still considered a Norwegian investment.[103]

No evidence has been provided to back up the widely circulated, if wildly exaggerated, claim by the U.S. think tank Center for Naval Analysis (CNA) that Chinese investments in Iceland from 2012 to 2017 were estimated at $1.25 billion (5.7% of GDP), including five major transactions averaging at $30.8 million.[104] On the contrary, no major Chinese investments took place in Iceland during this period or after. In another instance of misinformation, it has been claimed that China was, in 2018, intent on buying two ports in Iceland.[105] Even if the Chinese harbored such plans, they were not presented to the Icelandic government, which would never have approved them. Chinese authorities have, however, permitted Icelandic investment projects in China, with Icelandic energy companies, such as Arctic Green Energy, being heavily involved in developing geothermal resources in the country.[106]

In the aftermath of the financial crash, politicians from all parties had entertained high hopes about natural resource extraction within Iceland's EEZ. It was Skarphéðinsson, who pushed the issue, as has been discussed; it was subsequently taken up by Prime Minister Sigmundur Davíð Gunnlaugsson, who headed the Center-Right government, which was formed after the heavy defeat of the Social Democratic Alliance and the Left Green Movement in the 2013 parliamentary elections. In October the same year, an agreement on oil exploration in the Dragon Area was reached between China's CNOOC (with a 60% stake), the Norwegian state-owned energy firm Petoro (25%), and the Icelandic company Eykon Energy (15%). Gunnlaugsson declared that an oil find in the Dragon Area could result in a turnaround in Iceland's public finances and raise its credit rating.[107] This was another indication of how post-crisis politics shaped Iceland's Arctic policies based on nebulous ideas about future material returns. Such optimism continued to be expressed in the following years despite a drop in world oil prices,[108] which made exploration or extraction in remote northern spaces far less likely. Yet, in Iceland, counter-voices could also be heard, expressing opposition to resource exploitation in areas where it was difficult to respond to environmental disasters such as oil spills.[109] And in a rebuke to Skarphéðinsson, his own Social Democratic Alliance approved a resolution at its 2015 party congress, which criticized the oil exploration plans within Iceland's EEZ, and which led to a party policy change in a clear victory for environmentalists.[110]

This turnaround underscored that after recovering from the financial crisis, Iceland had far less need for the instrumentalization of the Arctic as part of hedging and crisis management agendas. Experiencing a tourist boom, it did not have to reach out for new economic opportunities or direct foreign investments from China or other countries. True, some of the Arctic projects, which were promoted by politicians after the banking collapse, continued. A German engineering company, Bremenports, for example, invested in preliminary research on the possibility of building a port in Finnafjörður in the northeast of Iceland[111] based on future transarctic route scenarios promoted by Skarphéðinsson. It subsequently reached an agreement with Icelandic partners to form a company for this aim, but so far nothing has come out of the project. Other plans were discarded, notably the oil exploration project in the Dragon Area. It was another sign that the strong revival of Iceland's economy had reduced the incentive for promoting overhyped Arctic investments. In 2018, it became clear that the foreign partners did not have enough faith in the feasibility of the Dragon project, with Petoro withdrawing from it and CNOOC following suit.[112] The Chinese decision, which did not square with a conspiracy-laden picture of China bent on making strategic long-term inroads into the Arctic, showed that business considerations overrode political ones in this case.

The Free Trade deal with China has proved important, but no game changer: From 2014 to 2023, exports to China tripled, with imports from it more than doubling. While the huge trade imbalance has favored China, its imports have stimulated Icelandic exports to China.[113] But only 2.2% of Iceland's exports went to China in 2023, while imports from China constituted 8.7% of total imports.[114] European countries and the United States remain Iceland's main trading partners. In 2023, 78% of Iceland's total exports went to the European Economic Area and the value of imports from it was 62% of the total.[115] In short, as a hedging exercise, the Chinese-Icelandic economic relationship followed a non-linear trajectory, with as many failures as accomplishments. Subsequently, it became, as we will see, even more, constrained due to renewed U.S. geopolitical and military interests in the North.

3.4 The Arctic Circle Assembly: Global Visions, Personal Legacies, and Nation Branding

When Iceland's President, Ólafur Ragnar Grímsson, launched, in mid-April 2013, the Arctic Circle Assembly initiative at the National Press Club in Washington, D.C., he claimed that he wanted to strengthen the policymaking process in the Arctic by bringing a wide range of stakeholders under one large tent.[116] What he had in mind was an annual conference, representing social as well as geographic inclusivity, with the participation of government representatives, political leaders, businesspeople, scientists, activists, and Indigenous peoples from the Arctic countries as well as non-Arctic ones from Asia, Europe, and other parts of the world. The Arctic Circle Assembly was a project that he and his collaborators, such as U.S. publishing executive Alice Rogoff, had been working on for some time prior to announcing it. In his presentation, Grímsson did not spell out a clear and developed Arctic agenda but stressed the effects of climate change and the melting of the sea ice in the region. Later, Grímsson argued that the "profound new significance of the Arctic in global terms" had required "a new platform on the scale of the World Economic Forum in Davos" with political leaders attending—or a "Medieval square where anyone could come knowing that they meet anybody and everybody there."[117]

The idea to create a broad-based and diverse public forum was reminiscent of his promotion of the Northern Research Forum, which had organized conferences in several Arctic countries. But the Arctic Circle initiative was far more ambitious and global in scope. It was described as an Arctic platform, with a "secretariat"—and, to add to the luster, the word "Assembly" was affixed to its name. What this suggested was that it had something to do with an international institution, which was not the case. In addition, the reference to Davos—as a symbolic site where the rich and powerful assembled—fitted uneasily with the inclusive rhetoric used to promote the undertaking or with the flaunting of "democratic" and "bottom-up" approaches. But since 2013, Grímsson has sought to mediate these paradoxical impulses at the annual Arctic Circle conferences by mixing elite panel sessions, where he is often the master of ceremony, with open academic ones without external interference, where scholars and researchers from universities and research institutes have presented their findings.

Following the launch, there was much international speculation about the motives behind the Arctic Circle initiative, which has become the largest annual gathering on Arctic affairs, with the number of participants increasing from 1200 in 2013 to over 2000 in 2023. Since Grímsson had stated that the Arctic lacked effective governance,[118] critical observers suggested that the Arctic Circle Assembly was meant to assume a specific political role outside the Arctic Eight and that it represented a challenge to the Arctic Council because of its governance aspirations and its "open door" policy vis-à-vis the non-Arctic states, especially those applying for admission to the Arctic Council as observers. At the National Press Club, Grímsson had, in fact, praised the Arctic Council as an "extraordinarily successful instrument" for Arctic cooperation. Still, since it was controlled by the Arctic states, he suggested that it lacked inclusiveness and that decisions had to be made on how it should "move forward."[119] More specifically, he offered the Arctic Circle Assembly as a venue for those who were not allowed to speak up at the Arctic Council, such as the observers.[120] To drive home the point, he cited France's Ambassador for the Polar Regions, Michel Rocard—a former Prime Minister—who claimed that the French were not used to being denied a voice at international forums. Unsurprisingly, Rocard addressed the first Arctic Circle conference later in the year.[121]

Grímsson's inclusive description of the Arctic Circle Assembly prompted Arctic scholars Duncan Depledge and Klaus Dodds to compare the gathering to a "bazaar" where the exchange of global and marginalized knowledge, ideas, and interests could influence Arctic governance.[122] What they termed "bazaar governance" is about how market power is exercised through informal dealings, networking, and hierarchies.[123] They did not argue that the Arctic Circle Assembly was competing with the Arctic Council for the right to govern Arctic affairs. Still, in a rather alarmist fashion, they interpreted the timing of Grímsson's initiative as a "provocation" to the Arctic Council, which was due to convene for its ministerial meeting in Sweden a month later to decide on the applications by Asian states and the EU Commission for an observer status. As they put it: "There was a palpable feeling at the time that Grímsson was challenging the Arctic Council to take on a more global profile, with a view that should the Arctic Council reject the observer applicants, then the Arctic Circle would be prepared to provide an alternative platform for global interests to be expressed in the Arctic."[124] In addition, they suggested that Arctic Circle could reposition Iceland, geopolitically, as a gateway for such non-conformist global Arctic perspectives.[125]

The establishment of the Arctic Circle Assembly had, in fact, nothing to do with the applications of non-Arctic states or the EU, or with pressuring the Arctic Council on this matter. Grímsson revived the old idea, which stemmed from the Reagan-Gorbachev Summit, of highlighting Iceland's geostrategic location for branding purposes. But he was not acting on behalf of the Icelandic government and promoting its Arctic policies. He rationalized the selection of Iceland as the first host of the Arctic Circle conference by saying that "after being ignored by much of the world," it had been put "on the global map" due to global warming.[126] Such rhetoric about Iceland's exceptionalism made little sense given that other places were far more affected by climate change. Depledge and Dodds theorized, however, that there was another political motive here: a China connection. They pointed out that Grímsson's announcement had coincided with the signing of the Free Trade agreement between Iceland and China and that two weeks later, the Icelandic firm Eykon Energy had teamed up with CNOOC, in the Arctic, to explore for oil in the Dragon Area. Furthermore, they claimed that the involvement of Alaskan and Greenlandic partners in the Arctic Circle Assembly betrayed a sub-state agenda, with autonomous Arctic voices that did not have to echo those of the U.S. and Danish governments.[127] Another commentator went much further, seeing the Arctic Circle Assembly, in ominous terms, as a "geopolitical interest of Grímsson's, in connecting subnational governments and non-Arctic states to the Arctic in a way that at some point has the potential to displace the Arctic Council and national Arctic sovereignty."[128] Yet another one made the outlandish point that the Arctic Circle Assembly—with a logo having six people holding hands—was a reaction to the exclusion of Iceland from the Ilulissat Declaration and the Arctic Five.[129] Thus, Grímsson's initiative was portrayed as a major geopolitical intervention aimed at making a direct stakeholding claim in Arctic governance by promoting the interests of non-Arctic states; at catapulting Iceland to a central global role in Arctic affairs as well as highlighting its political and economic interests in the region through its "blossoming relationship with China";[130] and at furthering a parastate agenda by creating a "network of the marginalized,"[131] which transcended sovereign interests of states in a potentially destructive way.

Such all-embracing and highly loaded political interpretations certainly gave the initiative the weight and mystic aura its initiator sought in the first place; in terms of status-seeking, the Arctic Circle Assembly soon became a major brand name, which the Icelandic government exploited. Still, most of these accounts were highly speculative, overblown, or plainly wrong. There was no validity to the suggestions that foreign state interests were behind the initiative or that a link existed between it and Iceland's Arctic policies, such as the coastal state claim. The same applied—as we have seen—to the decision by the Arctic Council in Kiruna to accept the six Asian states as observers. While Grímsson praised China's Arctic policy as being "constructive," his Arctic Circle announcement was totally unrelated to the Icelandic government's signing of the trade deal with China, the conclusion of the agreement with CNOOC, or other Arctic policy declarations or actions. And in contrast to big UN conferences, where political pressures can change state policies, there is no evidence to show that the Arctic Circle conference has had such impact on the Arctic policies of individual governments.[132]

Predictably, the Norwegian government was not happy with the launching of the Arctic Circle Assembly, seeing it as a competitor to the Arctic Frontiers conference in Tromsø, which has been held annually since 2007, and which has also attracted politicians, businesspeople, academics, and representatives from local Arctic communities. The Arctic Circle initiative also flew in the face of Norway's philosophy behind the Arctic Frontiers, which was about protecting the Arctic governing structure led by the Arctic states. As Norway's Foreign Minister, Espen Barth Eide put it: "We are happy that more people want to join our club [the Arctic Council as observers], because this means that they are not starting another club, and that gives us some influence on what topics are discussed in relation to the Arctic."[133]

In contrast to its Norwegian counterpart, which plays an instrumental role in formulating the program of the Arctic Frontiers and in inviting high-level participants, the Icelandic Foreign Ministry assumed no such function with respect to the Arctic Circle Assembly. In her study on Arctic conferences and governance, Beate Steinveg argues that the partnership between Arctic Circle and the Icelandic government supported "the principle of Iceland's 2011 Arctic policy" by advancing "Icelanders' knowledge of and discussion of the Arctic region."[134] While in a general sense, academics as well as officials have taken active part in the conference from the start, the Arctic Circle Assembly did not become a part of Iceland's Arctic policy until it was revised in 2021. It is true that many Icelandic government employees attend the conference. Yet the government's own involvement in the project is mostly about political symbolism, and it did not begin providing the annual gathering with modest financial and administrative support until several years after its commencement.

Grímsson's central public role in the Arctic Circle project was widely recognized abroad, but attempts to tie it to external interests were off the mark. When embarking on the initiative, Grímsson insisted—as several scholars have pointed out[135]—on viewing the Arctic in global terms rather than in regional ones where the Arctic states reign supreme. There was a need for involving non-Arctic states—including China, India, Nepal, and Pakistan—that were influenced by changes in the Arctic, such as melting of the sea ice and the Greenland glaciers, as well as in Antarctica and in the region identified with the "Third Pole"—the Himalayas, the ice-covered parts of Asia. Thus, the Arctic Circle Assembly was a venue, where the representatives of the Arctic Council observer states as well states with no such status could meet on an equal basis,[136] and where regional and sub-state interests and perspectives, such as those of Greenland, Alaska, and Scotland, could be addressed. An argument could certainly be made for transcending a self-serving hegemonial framework established by the Arctic states on the grounds that the social and ecological effects of climate change did not respect territorial boundaries or sovereign spaces. Such a catch-all approach, however, could also reproduce uncritical tendencies in climate change narratives, not to assign specific responsibility or single out culprits for global warming.[137]

Iceland's Arctic policy was far more restrictive, with its emphasis on UNCLOS and the Arctic Council—run by the Arctic Eight—as cornerstones of Arctic governance. Indeed, the Icelandic government's position had, in fact, much in common with that of the Norwegian government despite its rejection of the legitimacy of the Arctic Five venue. It was not consistent with Grímsson's argument that the Arctic Circle Assembly was "within the framework of Arctic governance" or "proof of a reasonable alternative model to the somewhat broken intergovernmental model."[138] It reflected the idealization of Arctic exceptionalism, where geopolitical divisions rooted in an East-West dichotomy had been relegated to a Cold War past. Grímsson had, as noted, strongly backed efforts to expand Iceland's ties with China following the financial crisis; he had been highly critical of the initial refusal of Western states to come to Iceland's rescue when its banking system collapsed, and he had, of course, suggested that Russia should be offered the former U.S. base. In addition, he publicly opposed Iceland's membership in the European Union, even after the Left-Wing government had started negotiations with it on joining the club.[139]

Hence, his political views sometimes clashed with government policies, which could lead to confusion and misunderstanding abroad. With the Icelandic President having no formal constitutional policymaking role, if a bully pulpit one, the officeholder can, in fact, only exert institutional power in limited circumstances: by deciding, if sometimes only symbolically, which political party leader receives the formal mandate to form a government after parliamentary elections, by having a say over the dissolution of parliament, and by refusing to sign laws passed by parliament. The non-ratification option does not entail a formal veto power but triggers an automatic national referendum on the legislation in question. Grímsson was the first President since the foundation of the Icelandic republic in 1944 to use this instrument to challenge the government and parliament. He did so in 2004, when he refused to sign a controversial law imposing restrictions on media ownership. Before the law was referred to a referendum, the government withdrew it.[140]

After the banking crash, he took an unambiguous stance against the two agreements negotiated by the Left-Wing government with the governments of the United Kingdom and the Netherlands over the Icesave accounts, in which Iceland accepted legal responsibility and monetary liabilities. Grímsson refused to ratify the laws underpinning the agreements, and Icelandic voters[141] overwhelmingly rejected them in two referenda in 2010 and 2011, respectively. Thus, Grímsson, whose political standing and reputation had been badly tarnished because of his cheerleading support for Iceland's oligarchic financial elites and the banking expansion abroad prior to the 2008 crash, managed to reinvent himself as a "people's champion" in the diplomatic disputes with the British and Dutch. Indeed, his popularity soared after siding with the anti-Icesave movement and by refusing to sign the Icesave bills, paving the way for a re-election to a fifth term in 2012, which he won decisively.[142] In a European Free Trade Association (EFTA) Court ruling in 2013, it was confirmed that there was no sovereign liability in relation to the Icesave accounts. Thus, the case for the non-payment of foreign debts incurred by the bankrupt Icelandic private bank was vindicated, with Grímsson among those reaping the political benefits.[143]

This drama played a part in creating the false impression abroad that in his capacity as President, Grímsson was in charge of Iceland's foreign policy, even if he had no such formal power or leverage. By taking a stance against the Left-Wing government—whether on EU accession, the Icesave issue, or constitutional reform—Grímsson alienated a sizeable group of his erstwhile leftist supporters and became, ironically, the darling of the Right, which had vilified him during most of his political career. Still, Skarphéðinsson—who had fought for Iceland's EU accession and who, as a Minister for Foreign Affairs, was implicated in the agreements with the British and the Dutch—praised him after leaving office in 2016 for testing the limits of the President's political power and for making the Arctic a part of his presidential portfolio. A former Left-Wing protégé of Grímsson, Skarphéðinsson gleefully pointed out that politicians, such as Jonas Gahr Støre, the head of the Norwegian Labor Party and, then, leader of the opposition, had accused Norway's conservative government of letting the Icelanders "steal" the Arctic.[144] Støre had, indeed, stated that he feared that Norway could lose its initiative on the Arctic, referring, specifically, to the presence of French President François Holland at the Arctic Circle Assembly.[145]

In contrast, shortly after leaving office in 2013, Skarphéðinsson had scolded Grímsson for claiming that some EU countries—without naming them—did not want to accept Iceland as a member state and for seeking to "take over Iceland's foreign policy."[146] During his tenure, Skarphéðinsson had defended what he termed President Grímsson's freedom of speech. Yet, even if Grímsson had used it to undercut the Left-Wing government's policies, government ministers had refrained from challenging him publicly, deferring to the symbolic presidential function as a unifier. Former Icelandic Foreign Ministers, such as Halldór Ásgrímsson (1995–2004) and Davíð Oddsson (2004–2005), had resisted President Grímsson's attempts to use the presidential office to promote his own views on foreign affairs.[147] His predecessors had interpreted and performed the President's role as a national reconciler, shunning political debates and controversies. It was, however, difficult for former political adversaries to see Grímsson in such a non-partisan light—and he did not hesitate to make political interventions. In the late 1990s, Ásgrímsson had been critical of Grímsson's dismissive comments on the EU as a supranational project in favor of European regional cooperation schemes such as the Arctic Council, the Council of the Baltic Sea States, and the Barents Euro-Arctic Council.[148] Oddsson even took the drastic step of cancelling scheduled meetings between Grímsson and U.S. officials of the executive branch, when the President visited Washington, D.C.[149] What Grímsson did to sidestep this diplomatic intervention was to meet with officials unofficially in restaurants and other venues.[150]

Despite their friendship, Skarphéðinsson's interest in the Arctic was not about parroting Grímsson's ideas about the region. On the contrary, Skarphéðinsson pushed, as has been shown here, his own Arctic initiatives, such as the cross-political 2011 Arctic parliamentary resolution, oil exploration in the Dragon Area, and a search and rescue center at the former U.S. military base. Thus, the Foreign Ministry played the primary role in developing and implementing Iceland's Arctic policies. The fear of losing control may partly explain why some foreign ministry officials had nothing but contempt for Grímsson's Arctic Circle Assembly project before the first conference was held.[151] Yet, after the successful launch, Icelandic government ministers and officials lavished the initiative with praise and supported it as a marketing and branding instrument.[152] Moreover, Icelandic prime ministers, foreign ministers, and other ministers as well as officials have taken part in the conference proceedings by giving speeches at plenary sessions, by taking part in roundtable discussions, by attending the conference without having any formal role in it, and/or by attending the Arctic Circle "Forums" abroad. Still, while the Arctic Circle has put international spotlight on Iceland and generated tourism-related economic benefits, it is far more identified with the person of Grímsson, who turned it, after a few years, largely into a one-man show.

Grímsson claimed that what underpinned the Arctic Circle Assembly was a grass-roots approach based on the notion that "formal representations of states no longer had the monopoly of the dialogue and where everyone could participate, whether he or she was a young activist, a student, a president or a prime minister."[153] In addition, he prided himself on never announcing a formal theme of the Arctic Circle conferences, for "[w]e create the platform, participants create the dialogue."[154] Yet such a characterization was, at best, incomplete. First, echoing Grímsson's own agenda, the plenary sessions are elitist and controlled in form and content. While his presence on the main conference stage was less conspicuous in the years when he was still President, he has since 2016 become the dominant figure, chairing the most high-profile sessions. Second, the absence of a formal conference agenda can be interpreted as showing a lack of focus, a complaint that has been expressed by some Arctic Circle participants, instead of being explained in "democratic" or non-hierarchical terms.

To be sure, the Arctic Circle Forums—which have been held in cooperation with governments, ministries, and organizations in host countries, including China, Korea, Japan, the United States, and the United Arab Emirates, and self-governing territories, such as Greenland, the Faroe Islands, and Scotland—have offered more targeted Arctic agendas. Research institutes such as the Sasakawa Peace Foundation, the Korea Polar Research Institute, and Polar Research Institute of China (PRIC) have been involved in organizing the Forums. Still, reflecting the Arctic Circle Assembly's heterogeneous and jumbled make-up, the academic Mia Bennet made the point in 2022 that it was "jokingly referred to as the Arctic Circus given the cast of characters who assembled in the opalescent, Olafur Eliasson-designed, concert hall, Harpa."[155] The building itself—situated by Reykjavík's harbor—has a loaded symbolic meaning of its own. It is inextricably tied to the ill-fated banking expansion of the mid-2000s when a leading Icelandic businessman, who later became bankrupt, spearheaded the construction project. Having started at the beginning of 2007, the work on Harpa was immediately halted after the 2008 financial collapse; for many, it became a testimony to the follies and excesses of that period, with suggestions made that it should be kept permanently in its half-finished state in commemoration of a dark chapter in Iceland's history.[156] After the Left-Wing government and the City of Reykjavík decided to finish the building, Harpa assumed yet another function following its inauguration in 2011: as a source of national pride—a modern monument of Iceland's expedite and successful reconstruction efforts. Thus, it became the perfect site and vehicle for a personal legacy and nation branding project such as the Arctic Circle Assembly.

In the first years, Grímsson was more successful in attracting high-profile participants to the event, when he could use the Office of the President in his invitation letters, blurring the line between the private character of the Arctic Circle Assembly as an NGO and his own public function. Thus, as mentioned, French President Hollande addressed the conference in 2015 and UN Secretary General Ban-ki Moon in 2016. Still, he has been able to attract ministers and former ministers from countries such as the United States, Finland, Greenland, Japan, and Scotland. He has also been keen on involving royalty in the proceedings—adding to their elitist flavor—by inviting the Swedish Crown Princess and the Norwegian Crown Prince to speak at the Arctic Circle Assembly. For several years, the rumor circulated that Russian President Vladimir Putin would attend the conference, but it did not materialize. In fact, few top Russian officials have participated in it despite Grímsson's wooing.[157] This is not to say that the Russians shunned the event; the Russian Academy of Science had, for example, a plenary session on Russian Arctic science at the 2018 conference, with the participation of high-level scientists. The Russians tended, however, to limit the attendance of high-ranking political representatives to their own Arctic conferences. An additional explanation could be that they did not like the inclusive vision of the global Arctic espoused by Grímsson; they took a narrow view of Arctic intergovernmental management relations, which should be restricted to the Arctic states and that non-Arctic states or supranational organizations, such as the European Union, should not be directly involved. As one senior Russian official put it, "the fewer, the better."[158] While the Russians eventually agreed to admitting the Asian states to the Arctic Council as observers, they were not among those states that pushed for increased external state or organizational representation. Whatever explains the rationale behind Russia's selection of participants in the Arctic Circle Assembly, there was nothing to suggest that it was reluctant to be associated with the project itself; after all, Chilingarov is an honorary board member and Anton Vasiliev, the former Senior Arctic Official and Ambassador to Iceland, an advisory board member.[159]

Consistent with his Arctic cooperative philosophy, Grímsson has always refrained from voicing criticism of Russia's foreign and Arctic policies. At an Arctic conference in Bodø in 2014, he responded to a high-level Norwegian official, who had found fault with Russia for annexing Crimea, by saying that this was not the right venue to express such criticisms. He made the point that the Arctic Council was the only post-Cold War forum for a Great Power dialogue and that it was important that the Arctic would not be divided up geopolitically.[160] When Russia invaded Ukraine in 2022, he neither blamed Russia nor Putin, whom characterized as a rather "modest" and "rational" person. Instead, he seemed to shift the blame on Western policy since the end of the Cold War, which he claimed had not succeeded in guaranteeing stability or peace in Europe. He also hinted at having an understanding for Putin's rationalization of the invasion by singling out NATO's Eastern expansion. And in a cryptic remark, which could be read as an apology for Russian behavior, he stated that a more "realistic" attitude toward Russia, which had no "democratic traditions," was "perhaps needed."[161] Yet, while stressing that there would be no winner in the war, he also sympathized with the plight of the Ukrainians and of the millions of refugees, lamenting the destruction of Ukraine's infrastructure. His ambiguous comments met with social media criticism in Iceland on the grounds that he was aligning with the aggressor, a charge he firmly denied.[162] Despite its pro-Ukrainian stance, the Icelandic political elite remained silent on the former President's intervention. Foreign Minister Þórdís Kolbrún Reykfjörð Gylfadóttir used, however, the opportunity at Grímsson's own Arctic Circle venue to criticize Russia in vocal terms, arguing that "we were experiencing the most dangerous times in decades as a consequence of Russia's illegal war against Ukraine and Putin's reckless threats that undermine regional and global security."[163]

When Grímsson launched the Arctic Circle Assembly, his positive views on China as an emerging global giant were nothing out of the ordinary. Except for Norway, whose relationship with China had been frozen in 2010 due to the Norwegian Nobel Committee's decision to award the Nobel Peace Prize to the Chinese dissident and imprisoned political activist Liu Xiaobo, there was, at that time, not much criticism of expanding Western political and economic relations with China. During the "diplomatic freeze," Norwegian officials sometimes chided their Icelandic colleagues for being too cozy to the Chinese, while, simultaneously, expecting Iceland's assistance in Norway's troubled relationship with China. Such a double-faced approach was quickly forgotten after the Norwegians made peace with the Chinese. After the formal normalization of Sino-Norwegian relations in 2016, both sides committed themselves to promoting "mutually beneficial" and, in a favorite Chinese parlor, "win-win cooperation" in various fields, including polar issues.[164] The first thing the Norwegian Prime Minister Erna Solberg did was to head a delegation of ministers and more than 200 Norwegian business leaders to Beijing to resume business as usual. This visit was followed up by another one in 2018, when Norway's Minister of Research and Education led a 250-strong delegation made up of the country's heads of universities and scientists to strengthen Chinese-Norwegian research collaboration.[165] After the Trump Administration reversed the Obama Administration's non-confrontational policy toward China by classifying it—and, to a lesser degree, Russia—as a global geo-strategic threat to the United States, the repercussions were quickly felt in other Western countries, including Iceland and Norway. Still, Grímsson has stuck to his inclusive vision of the Arctic, where both Russia as a Great Power and as a core Arctic state and China as a global power have a place.

This broad approach is reflected in the institutional setup of the Arctic Circle project. Its legal and administrative framework—from 2014 to 2021—was the Climate Research Foundation—a non-profit organization chaired by Dagfinnur Sveinbjörnsson, the CEO of Arctic Circle for the first eight years.[166] The purpose of the foundation was to promote international collaboration and scientific research on climate change. In 2021, the Arctic Circle Foundation took over the role, with Grímsson acting as its chair.[167] Apart from the secretariat, which is quite small and tasked with organizing the annual Assembly and the Forums, there is an honorary board as well as an advisory one. The honorary board only has a symbolic function and is made up of Grímsson, Prince Albert II of Monaco, U.S. Senator Lisa Murkowski, Sultan Ahmed Al Jaber, the United Arab Emirates' first Minister of Industry and Advanced Technology and Group CEO of the Abu Dhabi National Oil Company, and Artur Chilingarov. Murkowski has been a strong voice on Arctic affairs in the United States, and Chilingarov has, of course, been closely associated with the region for a long time, even if his current influence is mostly nominal. Prince Albert and Jaber are known internationally, if also for their engagement on other than Arctic issues. Controversially, Jaber, as head of one of the biggest oil producing companies in the world, presided over the UN climate change talks in Dubai (COP28) in 2023.[168] Given Jaber's involvement in the Arctic Circle project, it is not surprising that Grímsson staunchly defended his UN role.[169] Advisory board meetings are held once a year in connection with the Arctic Circle conference. The board is mostly composed of politicians, businessmen, officials, academics, and media people, such as Sara Olsvig, Chair of the Inuit Circumpolar Council; Alice Rogoff, the publisher of *Arctic Today*; Kuupik Kleist, former Prime Minister of Greenland; Huigen Yang, the former Director of the Polar Research Institute of China; and Skarphéðinsson.[170]

There has been much speculation about the financing of the Arctic Circle, which is registered as an NGO, with false suggestions made about the backing of China and Russia. According to financial statements, there is no evidence of financial support from either state.[171] In the early years, the main financial backers were private philanthropies and corporations, primarily in the United States, including the Carnegie Corporation of New York, MacArthur Foundation, Rockefeller Brothers Fund, Bloomberg Philanthropies, and Guggenheim Partners. Outside the United States, Masdar, the state-owned Abu Dhabi Future Energy Company, and, later, the UAE Ministry of Climate Change and Environment provided substantial funding. In addition, the Mamont Foundation and Prince Albert of Monaco Foundation have supported the initiative. The Icelandic government has provided a small annual grant—whose amount is lower than that spent on high-profile government-sponsored international conferences; it has also facilitated the attendance of Icelandic officials from various ministries at the conference.[172] Icelandic private companies have also given contributions. In recent years, an ever more sizeable portion of the conference funding is sourced from registrations fees.[173] The annual operating budget of the Arctic Circle Foundation amounts to about $1.7 million, which covers all its activities.[174]

Beate Steinveg has argued that the Arctic Circle Assembly and the Norwegian Arctic Frontiers Conference represent a new element of the soft-law dimension "of the Arctic governance architecture" and as part of a regime complex, "operating in the intersection between sovereign states and formalized cooperative arrangements."[175] Grímsson has echoed such a view by arguing that the Arctic Circle has added three components to the Arctic governance structure: by offering governments an international platform to present their views and politics; by giving subnational and regional entities a chance to advance their interests and perspectives independently of their central governments; and by bringing—through the Forums—attention to Arctic dialogues and cooperation outside the region, which has contributed to making non-Arctic states and entities constructive partners.[176] As a global brand, the Arctic Circle Assembly and

Forums have certainly become visible meeting places for government delegations, business interests, NGOs, Indigenous peoples, and academia. But as a high-profile venue for information sharing and networking, it should not be conflated with "Arctic governance"—as manifested in contractual relationships, institutional arrangements, and the pursuit of sovereign state interests—or cited as proof of its direct impact on Arctic state policies.

The periodic global media attention on the Arctic has not centered on stage-managed state promotions at the Arctic Circle but on concrete policy announcements or actions by individual Arctic states, such as the Ilulissat Declaration or the Russian North Pole flag-planting; territorial claims based on UNCLOS; UN reports on climate change or ice-melting; Arctic Council agreements on search and rescue, pollution prevention, and science cooperation; or statements issued by the United States, Russia, and other states, as well as NATO on the militarization of the Arctic. The Arctic Circle Assembly is primarily a performative platform with an international profile. Its philosophy is not dictated by the interests of other states or based on Iceland's foreign and Arctic policies, but, again, primarily reflects a global agenda as projected and articulated by its founder.

Notes

1. See, for example, Carsten Valgreen et al., *Iceland: Geyser crisis* (Copenhagen: Danske Bank, 2006), accessed February 20, 2024, https://www.mbl.is/media/98/398.pdf.
2. Kerry Capell, "The Stunning Collapse of Iceland," *NBC*, October 9, 2008, accessed February 18, 2023, https://www.nbcnews.com/id/wbna27104617.
3. See, for example, Valur Ingimundarson, *The Rebellious Ally: Iceland, the United States, and the Politics of Empire, 1945–2006*, 74, 181.
4. See Icelandic Prime Minister's Office, report, *The Economic Impact of the Russian Counter-Sanctions on Trade between Iceland and the Russian Federation*, January 2016, accessed February 2023, https://www.stjornarradid.is/media/forsaetisraduneyti-media/media/skyrslur/theeconomicimpactoftherussiansanctionsontradebetweenicelandandrussia.pdf.
5. Interview with a senior Russian official, June 22, 2009.
6. "Iceland seeks loan from Russia," *Bloomberg*, October 7, 2023, accessed June 3, 2023, https://www.business-standard.com/article/finance/iceland-seeks-loan-from-russia-108100801025_1.html.

7. “Davíð Oddsson: ‘Öll aðstoð frá Rússlandi er vel þegin’” [All Assistance from Russia Welcome], *Viðskiptablaðið*, September 6, 2010 [2008], accessedMay3,2023,https://www.vb.is/frettir/davi-oddsson-oll-asto-fra-russlandi-er-vel-egin/.
8. Interview with a senior Russian official, June 22, 2009.
9. Interview with a senior Russian official, June 22, 2009.
10. See, for example, Yuri Zarakhovich, “Why Russia Is Bailing Out Iceland,” *Time*, October 13, 2008, accessed February 18, 2023, https://content.time.com/time/world/article/0,8599,1849705,00.html.
11. Interview with an ambassador who was present at the meeting, June 18, 2009.
12. It was the Norwegian newspaper *Klassekampen* that first published the memorandum; see “Inviterar Russland” [Invites Russia], *Klassekampen*, November 12, 2008, accessed May 15, 2023, https://klassekampen.no/utgave/2008-11-12/inviterer-russland.
13. Interview with an ambassador who was present at the meeting, June 18, 2009.
14. A Report “Foreign Aid—Size and Composition,” no date [1965] Dennis Fitzgerald Papers, 1945–69, Box 44, Dwight D. Eisenhower Presidential Library, Abilene, Kansas.
15. See Valur Ingimundarson, “Buttressing the West in the North: The Atlantic Alliance, Economic Warfare, and the Soviet Challenge in Iceland, 1956–59,” *The International History Review* 21, no. 1 (1999): 80–103.
16. Interview with a high-ranking Icelandic official, November 15, 2008. See also Catherine Belton and Tom Braithwaite, “Iceland asks Russia for €4bn loan after west refuses to help,” *Financial Times*, October 8, 2008.
17. Interview with a high-level Russian official, June 22, 2009.
18. See “Samstarfsyfirlýsing ríkisstjórnar 2009” [The Platform of the Coalition Government [between the Social Democratic Alliance and the Left Greens] 2009], May 10, 2009, accessed February 22, 2024, https://www.stjornarradid.is/rikisstjorn/sogulegt-efni/um-rikisstjorn/2009/05/10/Samstarfsyfirlysing-rikisstjornar-2009/.
19. See “Send verður inn umsókn um aðild að ESB” [An EU Application will be Submitted], *Morgunblaðið*, July 16, 2009, accessed February 20, 2024, https://www.mbl.is/frettir/innlent/2009/07/16/samthykkt_ad_senda_inn_umsokn/.
20. See a summary of the Icesave case in Eiríkur Bergmann, “The Icesave Dispute: A Case Study into the Crisis of Diplomacy during the Credit Crunch,” *Nordicum Mediterraneum* 12, no. 1 (2017), accessed February 18, 2024, https://nome.unak.is/wordpress/volume-12-no-1-2017/double-blind-peer-reviewed-article/icesave-dispute-case-study-crisis-diplomacy-credit-crunch/; see also, for example, Ásgeir Jónsson and

Hersir Sigurgeirsson, *The Icelandic Financial Crisis: A Study into the World's Smallest Currency Area and its Recovery from Total Banking Collapse*; Valur Ingimundarson, Philippe Urfalino, and Irma Erlingsdóttir, *Iceland's Financial Crisis: The Politics of Blame, Protest, and Reconstruction*; Guðrún Johnsen, *Bringing Down the Banking System: Lessons from Iceland*.

21. See "Umsögn [orkumálastofnunar] um tillögur olíu og orkumálaráðuneytis Noregs varðandi Jan Mayen svæðið frá 12.3 1992" [Comment by the Icelandic National Energy Authority on the Proposals from Norway's Ministry of Oil and Energy from March 12, 1992], March 30, 1992, Folder, B/807, B-bréfasafn, 1991–1992, iðnaðar-og viðskiptaráðuneytið [Ministry of Industry and Commerce] 2013, ÞÍ.
22. See report, *Olíuleit á Drekasvæði við Jan Mayen-hrygg*, March 2007, Folder, F/0026, Skýrslur 1999–2009, iðnaðar- og viðskiptaráðuneytið, ÞÍ; see also memorandum (Kristján Skarphéðinsson and Pétur Örn Sverrisson) to Þorkell Helgason (Director General of the National Energy Authority), February 9, 2006, Folder, B/1319, B-bréfasafn, 2004–2007, iðnaðar og viðskiptaráðuneytið, ÞÍ.
23. Icelandic Ministry for Foreign Affairs, *Ísland á norðurslóðum* [Iceland in the Arctic], April 2009, accessed June 28, 2023, 7. https://www.utanrikisraduneyti.is/media/Skyrslur/Skyrslan_Island_a_nordurslodumm.pdf.
24. *Ísland á norðurslóðum*, 45–47.
25. "Iceland upset by Arctic summit snub," *CBC* News, February 16, 2010, accessed July 30, 2023, https://www.cbc.ca/news/canada/north/iceland-upset-by-arctic-summit-snub-1.885441; see also "Clinton til varnar Íslandi" "[[Hilary] Clinton Defends Iceland]," *RUV*, March 30, 2010, accessed February 10, 2024, https://www.ruv.is/frettir/innlent/clinton-til-varnar-islandi.
26. "Clinton rebuke overshadows Canada's Arctic meeting," *Reuters*, March 30, 2010, accessed May 20, 2023, https://www.reuters.com/article/us-arctic-idUSTRE62S4ZP20100330.
27. Eric J. Molenaar, "Participation in the Arctic Ocean Fisheries Agreement, in *Emerging Arctic Legal Orders*," ed. Akiho Shibata, Leilei Zou, Nikolas Sellheim, and Marzia Scopelli (Abingdon and New York, Routledge, 2019), 132–170.
28. See *Þingsályktun um stefnu Íslands í málefnum norðurslóða*. To put Iceland's Arctic policy within broader international context, the author—on the request of the Icelandic Foreign Ministry—wrote an analysis of Arctic geopolitics and governance, which accompanied the parliamentary resolution.

29. See Alyson Bailes and Lassi Heininen, *Strategy Papers on the Arctic or High North: a comparative study and analysis.*
30. Email communication from a former government minister, December 22, 2014.
31. *North Meets North: Navigation and the Future of the Arctic.*
32. See, for example, "Ræða Össurar Skarphéðinssonar utanríkisráðherra um utanríkismál" á Alþingi [Parliamentary Address by Foreign Minister Össur Skarphéðinsson], *Þingtíðindi*, September 14, 2010, accessed May 20, 2023, https://www.althingi.is/altext/raeda/138/rad20100514T105208.html.
33. "Ræða Össur Skarphéðinssonar um utanríkismál," May 14, 2010.
34. *Þingsályktun um stefnu Íslands í málefnum norðurslóða.*
35. See Egill Þór Níelsson, *The West Nordic Council in the Global Arctic* (Reykjavík: Institute of International Affairs, 2013).
36. *Þingsályktun um stefnu Íslands í málefnum norðurslóða.*
37. *Þingsályktun um stefnu Íslands í málefnum norðurslóða.*
38. *Þingsályktun um stefnu Íslands í málefnum norðurslóða.*
39. See "Ísland og Grænland" [Iceland and Greenland], memorandum on the Icelandic government's reaction to arguments in favor of making a territorial claim to Greenland [no date], n. d, 1993, 56, sögusafn utanríkisráðumeytis, ÞÍ; see also *Alþingistíðindi* [Parliamentary Records], 1954, D, umræður um þingsályktunartillögur og fyrirspurnir [Parliamentary Debates and Questions] (Reykjavík: Gutenberg Printing House, 1956); þingskjöl [Parliamentary Documents] 1954, A (Reykjavík: Gutenberg Printing House, 1955); Erik Beukel, Frede P. Jensen, and Jens Elo Rytter, *Phasing Out the Colonial Status of Greenland, 1945–55: A Historical Study* (Copenhagen: Museum Tusculanum Press, 2010); Daníel Hólmar Hauksson, "Grænlandsdraumurinn. Hugmyndir um tilkall Íslendinga til Grænlands á 20. öld" [The Greenland Dream: Ideas about Icelandic Claims to Greenland] (BA thesis, University of Iceland, 2019).
40. The Icelandic government was asked by the Norwegian government to coordinate a formal recognition of Kosovo a bit later than the other Nordic countries out of Norwegian deference toward Serbia, with which it had forged a historical relationship dating from World War II due to the presence of Yugoslav prisoners of war in Norway under Nazi occupation. Icelandic businessmen also pressed the government to go slow on recognition because they feared that it would detrimentally affect their commercial interests in Serbia; interview with a government minister, July 2, 2014.
41. See *Þingsályktun um stefnu Íslands í málefnum norðurslóða.*
42. See *Þingsályktun um stefnu Íslands í málefnum norðurslóða.*

43. See *Þingsályktun um stefnu Íslands í málefnum norðurslóða.*
44. See *Þingsályktun um stefnu Íslands í málefnum norðurslóða.*
45. See Dodds and Ingimundarson, "Territorial nationalism and Arctic geopolitics: Iceland as an Arctic Coastal State."
46. See Michael Billig, *Banal Nationalism* (London: Sage Publications).
47. *North Meets North: Navigation and the Future of the Arctic.*
48. See Valur Ingimundarson, "'A Crisis of Affluence': the Politics of an Economic Breakdown in Iceland," *Irish Studies in International Affairs*, 21 (2010): 57–69.
49. Styrmir Gunnarsson, *Umsátrið—fall Íslands og endurreisn* [Under Siege: The Collapse of Iceland and Its Reconstruction] (Reykjavík: Veröld), 71.
50. Interview with a former Senior Arctic Official, July 18, 2023.
51. Interview with a former Senior Arctic Official, July 18, 2023; interview with a high-ranking former Arctic Council official, February 19, 2024.
52. Rasmus Brun Pedersen, "Bandwagon for Status: Changing Patterns in Nordic States Status-seeking Strategies," 217–241; see also De Carvalho and Neumann, "Small states and status," 56–72.
53. Icelandic Ministry for Foreign Affairs, Press Release, "Launch of Icelandic Arctic Cooperation Network and arrival of first Nansen Professor," February 12, 2013, accessed June 25, 2023, https://www.government.is/news/article/2013/02/12/Launch-of-Icelandic-Arctic-Cooperation-Network-and-arrival-of-first-Nansen-Professor/.
54. See Arctic Council Secretariat, *Annual Report 2013*, March 2014, accessed November 23, 2023, https://oaarchive.arctic council.org/bitstream/handle/11374/941/ACS_Annual_Report_2013_as_printed.pdf?sequence=1&isAllowed=y.
55. Interview with a Senior Arctic Official, December 3, 2012; interview with an Icelandic foreign ministry official, January 13, 2023. After serving as the Russian SAO at the Arctic Council, 2008–2014, he was Russia's Ambassador to Iceland from 2014 to 2020.
56. Interview with a former high-level Arctic official, February 19, 2024.
57. "Russia hands over presidency of Arctic Council's Scientific Work to Norway," *TASS*, April 14, 2023, accessed February 10, 2023, https://tass.com/russia/1604697; see Anton Vasiliev, "A Story of an Image: The Arctic Council at 25—Reflections, Arctic Circle," Arctic Circle, September 20, 2021, accessed February 20, 2024, https://www.arcticcircle.org/journal/by-anton-vasiliev-russias-senior-arctic-official.
58. See "The Arctic as Global Challenge—Issues and Solution."
59. See *Skýrsla Össurar Skarphéðinssonar utanríkisráðherra um utanríkis- og alþjóðamál* [Report by Foreign Minister Össur Skarphéðinsson on Foreign and International Affairs], *Þingtíðindi*, April 2012, accessed September 30, 2023. https://www.althingi.is/altext/140/s/1229.

html; *Skýrsla Gunnars Braga Sveinssonar utanríkisráðherra um utanríkis- og alþjóðamál* [Report by Foreign Minister Gunnar Bragi Sveinsson on Foreign and International Affairs], *Þingtíðindi*, March 2014, accessed September 30, 2023, https://www.althingi.is/altext/143/s/0757.html.

60. See *Skýrsla Össurar Skarphéðinssonar utanríkisráðherra um utanríkis- og alþjóðamál Þingtíðindi*, April 2012.
61. U.S. Department of Defense, *Arctic Strategy*, November 2013, accessed September 30, 2022, 13, https://www.hsdl.org/?abstract&did=747036.
62. See, for example, Össur Skarphéðinsson, "Viðbragðsmistöð á norðurslóð" [A Rescue Center in the North], *Morgunblaðið*, March 14, 2013, accessed January 16, 2023, https://www.mbl.is/greinasafn/grein/1458610/.
63. See Gunnar Bragi Sveinsson, "Iceland's Role in the Arctic—The Future of Arctic Cooperation," Arctic Circle Assembly Reykjavík, October 14, 2013, accessed November 15, 2022, https://www.stjornarradid.is/media/utanrikisraduneyti-media/media/nordurslodir/Arctic-Circle-speech-October-14-2013.pdf, see also Sveinsson, "Iceland's Policy and Priorities in a Changing Arctic," January 20, 2014, accessed September 30, 2022, https://www.stjornarradid.is/media/utanrikisraduneyti-media/media/Raedur/Raeda-GBS-ArcticFrontiers.pdf.
64. Interviews with Icelandic officials, June 13, 2016.
65. Ciorciari and Haacke, "Hedging in international relations," 372.
66. See Darren J. Lim and Rohan Mukherjee, "Hedging in South Asia: balancing economic and security interests amid Sino-Indian competition," *International Relations of the Asia-Pacific* 19, no. 3 (2019): 493–522.
67. Örn Daníel Jónsson, Ingjaldur Hannibalsson, and Li Yang, "A bilateral free trade agreement between Iceland and China," in *Þjóðarspegillinn—Rannsóknir í félagsvísindum*, ed. Ingjaldur Hannibalsson (Reykjavík: Félagsvísindastofnun Háskóla Íslands, 2013), 1–7, accessed January 27, 2023, https://skemman.is/en/stream/get/1946/16786/39049/3/OrnDJonsson_VID.pdf; see also Baldur Thorhallsson and Snæfridur Grimsdottir, *Lilliputian Encounters with Gulliver: Sino–Icelandic Relations from 1995 to 2021* (Reykjavík: International Institute at the University of Iceland, 2021).
68. The agreement was renewed in 2013 and 2016, but not in 2019, probably because of the strong position of Iceland's currency reserves.
69. See, for example, Jóhanna Sigurðardóttir, "Beðist afsökunar á vanrækslu og andvaraleysi" [An Apology for Failures and Complacency], Prime Minister's Office, October 6, 2009, accessed February 20, 2024, https://www.stjornarradid.is/efst-a-baugi/frettir/stok-frett/2009/10/06/Bedist-afsokunar-a-vanraekslu-og-andvaraleysi/.

70. See, for example, "Iceland invites China to Arctic shipping," *The Barents Sea Observer*, September 22, 2010, accessed February 23, 2023, https://barentsobserver.com/en/sections/business/iceland-invites-china-arctic-shipping.
71. "Ólafur Ragnar kominn til Kína," *Vísir,* May 16, 2005, accessed February 10, 2024, https://www.visir.is/g/20051182278d; "Ræddi meðal annars mannréttindamál við forseta Kína" [Discussed Human Rights Among Other Things with the President of China], *Morgunblaðið*, May 17, 2005, accessed February 10, 2024, https://www.mbl.is/frettir/innlent/2005/05/17/raeddi_medal_annars_mannrettindamal_vid_forseta_kin/; Kári Jónasson, "Árangur af Kínaheimsókn forsetans," *Vísir*, May 20, 2006, accessed February 10, 2024, https://varnish-7.visir.is/g/2005505200301/arangur-af-kinaheimsokn-forsetans.
72. See, for example, Andrew Stephen Campion, "From CNOOC to Huawei: securitization, the China Threat, and critical infrastructure," *Asian Journal of Political Science*, 29, no 1 (2020): 47–66.
73. It was called a "Framework Agreement on Arctic cooperation and Memorandum of Understanding (MOU) on Cooperation in the Field of Marine and Polar Science and Technology"; see Embassy of Iceland in Beijing, "China and Iceland sign agreements on geothermal and geoscience cooperation and in the field of polar affairs," April 23, 2012, accessed February 23, 2023, https://www.iceland.is/iceland-abroad/cn/english/news-and-events/china-and-iceland-sign-agreements-on-geothermal.
74. See Egill Þór Níelsson, "China Nordic Arctic Research Center," in *Nordic-China Cooperation: Challenges and Opportunities*, ed. Andreas Bøje Forsby (Copenhagen: NIAS Press, 2019), 59–63; see also Shan Yanyan, Jianfeng He, Guo Peiqing, and Liu He, "An assessment of China's participation in polar subregional organizations," *Advances in Polar Science*, 34, no. 1 (2023): 56–65.
75. Icelandic Ministry for Foreign Affairs. "Free Trade Agreement between Iceland and China. Fact Sheet," April 15, 2013, accessed February 23, 2023, https://www.government.is/topics/foreign-affairs/external-trade/free-trade-agreements/free-trade-agreement-between-iceland-and-china/.
76. Icelandic Ministry for Foreign Affairs, *Ísland á norðurslóðum.*
77. See Össur Skarphéðinsson, *Ár drekans. Dagbók utanríkisráðherra á umbrotatímum* [The Year of the Dragon: Diaries of a Foreign Minister in Turbulent Times] (Reykjavík: Sögur, 2013), 48, 53–53.
78. See Skarphéðinsson, *Ár drekans*, 52–53.
79. Skarphéðinsson, *Ár drekans*, 52–53.

80. See Serafettin Yilmas, "Exploring China's Arctic Strategy: Opportunities and Challenges," *China Quarterly of International Strategic Studies* 3, no. 1 (2017): 57–78; Mariia Kobzeva, "Strategic partnership setting for Sino-Russian cooperation in Arctic shipping," *The Polar Journal 10*, no. 2 (2020): 334–352.
81. Atle Staalesen, "Chinese shippers shun Russian Arctic waters," *The Barents Observers*, August 22, 2022, accessed February 20, 2024, https://thebarentsobserver.com/en/industry-and-energy/2022/08/chinese-shippers-shun-russian-arctic-waters.
82. The Newnew company quickly came under a cloud when in October 2023, one of its container ships, Polar Bear, was suspected of damaging two underwater telecommunications cables—between Estonia and Sweden and Estonia and Finland—and a natural gas pipeline in the Baltic Sea. The Russians subsequently stated that one of its own telecommunications cable had also been damaged by the Polar Bear, whose anchor had been dislodged. The Finnish authorities stated that they thought that the act had been intentional. The Polar Bear subsequently returned to China via the Northern Sea Route under the escort of Russian icebreakers. See "Estonia Says Chinese Ship is Main Focus of Probe Into Cables Damage," *Reuters*, November 10, 2023, accessed February 25, 2024, https://gcaptain.com/estonia-eyes-chinese-ship-cable-damage-probe/; "Russian Firm Says Baltic Telecoms Cable was Severed as Chinese Ship Passed Over," *Marine Link*, November 7, 2023, accessed February 25, 2024, https://www.marinelink.com/news/russian-firm-says-baltic-telecoms-cable-509285; "Finland says vessel's broken anchor caused Balticconnector damage," *Aljazeera*, October 25, 2023, accessed February 25. 2024, https://www.aljazeera.com/news/2023/10/25/finland-pipeline-damage-cause; "'Everything indicates' Chinese ship damaged Baltic pipeline on purpose, Finland says," *Politico*, December 1, 2023, accessed February 25, 2024, https://www.politico.eu/article/balticconnector-damage-likely-to-be-intentional-finnish-minister-says-china-estonia/; Atle Staalesen, "Newnew Polar Bear sails towards Bering Straits," *The Barents Observer*, November 6, 2023, accessed February 24, 2024, https://thebarentsobserver.com/en/security/2023/11/newnew-polar-bear-exits-northern-sea-route.
83. On China's Arctic aspirations at that time, see Linda Jakobson and Jingchao Peng, *China's Arctic Aspiration*. SIPRI Policy Paper no. 34, Stockholm: International Peace Research Institute, 2012; Kai Sun, "China and the Arctic: China's Interests and Participation in the Region," *East Asia-Arctic Relations: Boundary, Security and International Politics*, Paper 2. GIGI (November 2013), accessed November 11, 2022, https://www.cigionline.org/publications/east-asia-arctic-relations-boundary-

security-and-international-politics/; Nong Hong, "Emerging interests of non-Arctic countries in the Arctic: A Chinese perspective," *Polar Journal* 4, no. 2 (2014): 271–286.

84. See Alexeeva and Lasserre, "The Snow Dragon: China's Strategies in the Arctic," *China Perspectives*, 3 (2012): 61–68.
85. See "CNOOC to lead offshore Iceland license group: Orkustofnun, Iceland's National Energy Authority has granted its third license in the offshore Dreki area," *Offshore*, February 6, 2014, accessed February 10, 2024, https://www.offshore-mag.com/regional-reports/article/16783755/cnooc-to-lead-offshore-iceland-license-group; *Ársskýrsla Orkustofnunar 2013* [The Annual Report of the Icelandic National Energy Authority] (Reykjavík: Orkustofnun 2014), accessed February 10, 2024, https://rafhladan.is/bitstream/handle/10802/18132/OS-arsskyrsla-2013.pdf?sequence=33; "Oil exploration in Icelandic waters comes to an end: Too expensive and too risky," *Iceland Magazine*, January 23, 2018, accessed February 10, 2024, https://icelandmag.is/article/oil-exploration-icelandic-waters-comes-end-too-expensive-and-too-risky.
86. Xavier-Bender, "China and the West May Soon Compete for Troubled Iceland," *RealClear World*, May 7, 2013, accessed July 5, 2023, https://www.realclearworld.com/articles/2013/05/07/china_and_the_west_may_soon_compete_for_troubled_iceland_105141.html.
87. The Arctic Institute, Center for Circumpolar Security Studies, "Iceland for Sale—Chinese Tycoon Seeks to Purchase 300 km^2 of Wilderness," September 2, 2011, accessed July 5, 2023, https://www.thearcticinstitute.org/iceland-sale-chinese-tycoon/.
88. Information supplied by the Icelandic Ministry for Foreign Affairs, May 25, 2012.
89. "ASÍ mótmælir fríverslunarsamningi við Kína" [The Icelandic Federation of Labor Protests the Free Trade Agreement with China], *Vísir*, April 12, 2013, accessed January 19, 2023, https://www.visir.is/asi-motmaelir-friverslunarsamningi-vid-kina/article/2013130419728. April 12, 2013.
90. On the Huang Nubo case, see the autobiography of former Minister of the Interior, Ögmundur Jónasson, *Rauði þráðurinn* (Selfoss: Bókaútgáfan Sæmundur, 2022), 411–424; interview with a government minister, April 10, 2013.
91. See "Teeing Off at Edge of the Arctic? A Chinese Plan Baffles Iceland," *New York Times*, March 22, 2013, accessed January 19, 2023, https://www.nytimes.com/2013/03/23/world/europe/iceland-baffled-by-chinese-plan-for-golf-resort.html.
92. See, for example, "Stjórn Húsavíkurstofu styður áform Huang Nubo" [The Tourist Board of the Húsavík Municality Support Huang Nubo's

Plans], *640.is*, November 22, 2011, accessed February 20, 2024, https://www.640.is/is/frettir/stjorn-husavikurstofu-stydur-aform-huang-nubo.

93. Jónasson, *Rauði þráðurinn*, 416–420; interview with a government minister, April 10, 2013.
94. See Jónasson, *Rauði þráðurinn*, 416.
95. "Hætta við kaup á Grímsstöðum" [Abandon Plans for the Purchase of Grímsstaðir], *RUV* [Icelandic Public Radio Broadcasting Services], December 12, 2014, accessed January 19, 2023, https://www.ruv.is/frett/haetta-vid-kaup-a-grimsstodum.
96. See, for example, "British tycoon buys vast farm in the highlands. Intends to do 'absolutely nothing with it,'" *Iceland Magazine,* December 16, 2016, accessed January 4, 2023, https://icelandmag.is/article/british-billionaire-buys-vast-farm-highlands-intends-do-absolutely-nothing-it.
97. See "Ögmundur segir sölu Grímsstaða mark um vesaldóm stjórnvalda" [Ögmundur [Jónasson] Characterizes the Sale of the Land of Grímsstaðir as an Example of the Government's Spinelessness], *Vísir*, December 20, 2016, accessed May 5, 2023, https://www.visir.is/g/2016161229918/ogmundur-segir-solu-grimsstada-mark-um-vesaldom-stjornvalda.
98. Interviews with Senior Arctic officials, December 12, 2012, and May 28, 2013.
99. Arctic Council Secretariat, *Kiruna Senior Arctic Officials Report to Minister*, May 15, 2013.
100. See Sveinn K. Einarsson, Ingjaldur Hannibalsson, and Alyson J.K. Bailes, "Chinese Investment and Icelandic National Security," in *Þjóðarspegillinn*, ed. Silja B. Ómarsdóttir (Reykjavík Félagsvísindastofnun Háskóla Íslands, 2014), 1–13, accessed January 21, 2023, https://www.semanticscholar.org/paper/Chinese-Investment-and-Icelandic-National-Security-Einarsson-Bailes/6251bacb83a9c6b6e24bd7b84b29a1d5973fe42c.
101. OECD, "OECD Benchmark Definition of Foreign Direct Investment—FOURTH EDITION," 2008, accessed May 13, 2022, https://www.oecd.org/daf/inv/investmentstatisticsandanalysis/40193734.pdf. P. 17.
102. Chinese Embassy in Iceland, "Sino-Icelandic Economic and Trade Relationship," Embassy of the People's Republic of China in the Republic of Iceland, 2019, accessed April 16, 2022, https://is.china-embassy.gov.cn/eng/zbgx/jmgx/201904/t20190410_3164334.htm.
103. Askja Energy, "Foreign Investment [in Iceland]," n.d. accessed June 20, 2023. https://askjaenergy.com/iceland-investing/foreign-investment/.
104. See Mark E. Rosen and Cara B. Thuringer, "Unconstrained Foreign Direct Investment: An Emerging Challenge to Arctic Security," CNA's Occasional Paper series (November 2017), accessed June 25, 2023, 54, https://www.cna.org/cna_files/pdf/COP-2017-U-015944-1Rev.pdf.

105. Nong Hong, *China's Role in the Arctic: Observing and Being Observed* (London, New York: Routledge, 2020), 151.
106. "Chinese Embassy in Iceland, Science and Technology Exchanges," April 9, 2019, accessed February 10, 2024, http://is.china-embassy.gov.cn/eng/zbgx/kjjl/201904/t20190409_3165020.htm.
107. "Trú á olíufund myndi strax bæta lánstraust Íslendinga" [A Belief in an Oil Find Would Immediately Improve Iceland's Credit Rating], *Fréttablaðið*, November 7, 2013, accessed January 16, 2023, https://timarit.is/page/6142786#page/n0/mode/2up.
108. See, for example, "Drekasvæðið mun betra en við þorðum að vona" [The Dragon Area Has Far More Potential Than We Had Dared to Hope], *Vísir*, October 11, 2016, accessed January 16, 2023, https://www.visir.is/g/2016161019705/-drekasvaedid-mun-betra-en-vid-thordum-ad-vona-.
109. See, for example, Jónas Haraldsson, "Olíuvinnslan vart áhættunnar virði" [Oil Extraction Hardly Worth the Risk], *Fréttatíminn*, November 7–9, 2014, accessed January 16, 2023, https://timarit.is/page/6194027#page/n12/mode/2up.
110. "Mistök að styðja olíuleit á Drekasvæðinu" [A Mistake to Support Oil Exploration in the Dragon Area], *RUV*, March 22, 2015, accessed January 16, 2023. https://www.ruv.is/frett/mistok-ad-stydja-oliuleit-a-drekasvaedi.
111. Bremensport, "Welcome to the Finnafjord Project," accessed February 20, 2024, https://bremen-ports.de/finnafjord/.
112. "Olíuleit í uppnámi" [Oil Exploration Plans in Limbo] *Viðskiptablaðið*, January 25, 2018, accessed January 16, 2023, https://www.vb.is/frettir/oliuleit-i-uppnami/.
113. See Helga Kristjánsdóttir, Sigurður Guðjónsson, and Guðmundur Kristján Óskarsson, "Free Trade Agreement (FTA) with China and Interaction between Exports and Imports," *Baltic Journal of Economic Studies*, 8, no. 1 (2022), DOI: https://doi.org/10.30525/2256-0742/2022-8-1-1-8.
114. *Skýrsla utanríkisráðherra um utanríkis- og alþjóðamál* [Report by the Minister for Foreign Affairs on Foreign and International Affairs], May 2024, 135, accessed October 25, 2024, https://www.stjornarradid.is/gogn/rit-og-skyrslur/stakt-rit/2024/05/14/Skyrsla-utanrikisradherra-um-utanrikis-og-althjodamal/.
115. Hagstofa Íslands [Statistics Iceland] ,"Vöruviðskipti óhagstæð um 363,3 milljarða árið 2023" [The Trade Imbalance Amounts to 363.3 billion Icelandic Kronur in 2023], accessed October 25, 2024, https://hagstofa.is/utgafur/frettasafn/utanrikisverslun/voruvidskipti-i-arid-2023-lokatolur/.

116. On the Arctic Circle Assembly see, for example, Beate Steinveg, "Governance by conference? Actors and agendas in Arctic politics," PhD diss. (The Arctic University of Norway, 2020); see also Steinveg, "The role of conferences within Arctic governance," *Polar Geography* 44, no. 1 (2021): 37–54; Steinveg, "Exponential Growth and New Agendas—A Comprehensive Review of the Arctic Conference Sphere," *Arctic Review on Law and Politics* 12 (2021): 134–160; Duncan Depledge and Klaus Dodds, "Bazaar Governance: Situating the Arctic Circle," in *Governing Arctic change: global perspectives*, ed. Kathrin Keil and Sebastian Knecht (London: Palgrave Macmillan, 2017), 141–160; Lara Johannsdottir and David Cook, "Discourse analysis of the 2013–2016 Arctic Circle Assembly programmes," *Polar Record* 53 (2017): 276–279.
117. Vilborg Einarsdóttir, "Ólafur Ragnar Grímsson: A New Model of Arctic Cooperation for the 21st Century," *JONAA, Journal of the North Atlantic & Arctic* (November 2018), accessed January 9, 2022, https://www.jonaa.org/content/2018/10/19/a-new-model.
118. See Robert Webb, "Iceland president sounds climate alarm demanding global attention, action at NPC Luncheon," April 15, 2015, accessed January 7, 2023, https://www.press.org/newsroom/iceland-president-sounds-climate-alarm-demanding-global-attention-action-npc-luncheon.
119. "Icelandic President Ólafur Ragnar Grímsson speaks at April 17, 2013 National Press Club Luncheon," *YouTube*, April 17, 2013, accessed May 25, 2023, https://www.youtube.com/watch?v=wW0p_Eh94PI.
120. "Icelandic President Ólafur Ragnar Grímsson speaks at April 17, 2013 National Press Club Luncheon."
121. Michel Rocard, Address at the Arctic Circle Assembly, *YouTube*, November 1, 2014, accessed June 23, 2023, https://www.youtube.com/watch?v=qleb5wprlhw&list=PLI0a77tmNMvRkshNIRBriGV-K8o3heehr&index=24.
122. See Depledge and Dodds, "Bazaar Governance: Situating the Arctic Circle," 143, 146.
123. On the concept, see, for example, Morteza Aalabaf-Sabaghi, "Bazaar Governance," *SSRN Electronic Journal* (February 27, 2008): 1–5, accessed January 18, 2023, https://doi.org/10.2139/ssrn.1098783.
124. See Depledge and Dodds, "Bazaar Governance: Situating the Arctic Circle," 142–143.
125. Depledge and Dodds, "Bazaar Governance: Situating the Arctic Circle," 143.
126. Webb, "Iceland president sounds climate alarm demanding global attention, action at NPC Luncheon."

127. Depledge and Dodds, "Bazaar Governance: Situating the Arctic Circle," 143–144.
128. Steinveg, "Governance by conference? Actors and agendas in Arctic politics," 150. The academic commentator is anonymous.
129. Steinveg, "Governance by conference? Actors and agendas in Arctic politics," 139. The academic is not named.
130. See Depledge and Dodds, "Bazaar Governance: Situating the Arctic Circle," 143.
131. See Depledge and Dodds, "Bazaar Governance: Situating the Arctic Circle," 143–144.
132. See Steinveg, "Governance by conference? Actors and agendas in Arctic politics," 160.
133. Espen Barth Eide, "Opening Remarks at Arctic Frontiers: The Arctic—The New Cross-roads," January 21, 2013, accessed February 21, 2024, https://www.regjeringen.no/en/historical-archive/Stoltenbergs-2nd-Government/Ministry-of-Foreign-Affairs/taler-og-artikler/2013/remarks_frontiers/id713462/; see also Steinveg, "Exponential Growth and New Agendas—A Comprehensive Review of the Arctic Conference Sphere," *Arctic Review on Law and Politics* 12 (2021): 151.
134. Steinveg, "Governance by conference? Actors and agendas in Arctic politics," 147.
135. See, for example, Depledge and Dodds, "Bazaar Governance: Situating the Arctic Circle," 141–161.
136. Vilborg Einarsdóttir, "Ólafur Ragnar Grímsson: A New Model of Arctic Cooperation for the 21st Century."
137. See Mia M. Bennett, "Rise of Sinocene: China as a Geological Agent," in *Infrastructure and the Remaking of Asia*, ed. Max Hirsh and Till Mostowlansky (Honolulu University of Hawai'i Press, 2020), 30.
138. See Steinveg, "Governance by conference? Actors and agendas in Arctic politics," 234.
139. See, for example, "Ísland í betri stöðu utan ESB" [Iceland in a Better Position Outside the EU], *Morgunblaðið*, December 23, 2012, accessed July 2, 2023, https://www.mbl.is/frettir/innlent/2012/12/13/island_i_betri_stodu_utan_esb/.
140. See, for example, "Staðfest að fjölmiðlalögin verði afturkölluð" [It Has Been Confirmed That the Media Law Will Be Withdrawn], *Morgunblaðið*, July 20, 2004, accessed June 15, 2023, https://www.mbl.is/frettir/innlent/2004/07/20/stadfest_ad_fjolmidlalog_verda_afturkollud/.
141. See, for example, Fahad Sayeed, "Sovereign default of Iceland: Voting outcomes of the referenda 2010 and 2011 conducted for the approval of the Icesave bill," *SSRN*, March 13, 2015, accessed February 20, 2024, https://papers.ssrn.com/sol3/papers.cfm?abstract_id=2571216.

142. Kosningasaga. Upplýsingar um kosningar á Íslandi [History of Elections: Electoral Information in Iceland], "Forsetakosningar 2012" [Presidential Elections], https://kosningasaga.wordpress.com/forsetakosningar/forsetakosningar-2012/.
143. See, Press Release, "Judgment in Case E-16/11 EFTA Surveillance Authority v Iceland ('Icesave'). Application of the EFTA Surveillance Authority in the Case of Icesave Dismissed," EFTA Court, January 28, 2013, accessed February 20, 2024, https://eftacourt.int/wp-content/uploads/2019/01/16_11_PR_EN1.pdf?x85212.
144. Össur Skarphéðinsson, "Auknar og breyttar kröfur á forseta" [Additional and Changed Demands on the President], *Fréttablaðið*, March 23, 2016, accessed January 16, 2023, https://timarit.is/page/6509020#page/n19/mode/2up.
145. "Støre frykter Norge kan miste nordområdeinitiativet" [Støre Fears that Norway Can Lose the Initiative on the Arctic], *NRK*, November 10, 2015, accessed January 19, 2022, https://www.nrk.no/tromsogfinnmark/store-frykter-norge-kan-miste-nordomradeinitiativet-1.12646936.
146. "Össur gagnrýnir Ólaf fyrir ummæli um ESB" [Össur Skarphéðinsson Criticizes Ólafur Ragnar Grímsson for His Comments on the EU], *RUV*, June 6, 2013, https://www.ruv.is/frettir/innlent/ossur-gagnrynir-olaf-fyrir-ummaeli-um-esb.
147. See, for example, a parliamentary speech by Halldór Ásgrímsson, "Orð forseta um Evrópusambandið" [The President's Comments on the European Union], *Þingtíðindi*, April 17, 2002, accessed January 16, 2023, https://www.althingi.is/altext/127/04/r17104201.sgml.
148. See "Halldór Ásgrímsson: Sjónarmið forseta og ríkisstjórnar fari saman" [The President and the Government Should Speak with One Voice], *Morgunblaðið*, November 25, 1998, accessed January 17, 2023, https://www.mbl.is/greinasafn/grein/433648/.
149. An interview with an Icelandic Foreign Ministry official, August 28, 2004.
150. See Ólafur Ragnar Grímsson, *Sögur handa Kára* [Stories for Kári [Stefánsson]] (Reykjavík: Mál og menning, 2020).
151. High-ranking Icelandic government official, interview by author, February 5, 2019.
152. See, for example, "Sigmundur: Mikið í húfi" [PM Sigmundur Davíð Gunnlaugsson: Much at Stake], *Morgunblaðið*, October 31, 2014, accessed February 15, 2024, https://www.mbl.is/frettir/innlent/2014/10/31/sigmundur_mikid_i_hufi/; "Katrín segir að Hringborð norðurslóða hafa breytt umræðunni" [PM Katrín Jakobsdóttir Says That the Arctic Circle Has Changed the Debate], *Vísir*, October 19, 2018, accessed February 15, 2024, https://www.visir.is/g/2018181018681/f/f/skodanir; see also comments by Foreign Minister Bjarni

Benediktsson, *Facebook,* October 20, 2023, accessed February 15, 2024, https://m.facebook.com/story.php?story_fbid=pfbid032vAy9Y41S8VY1QAmmRK8xAobZPXQa8hJR64SgbqowC3b6K4k5UdHf2PaWpZkmDPCl&id=100044612311463.

153. Vilborg Einarsdóttir, "Ólafur Ragnar Grímsson: A New Model of Arctic Cooperation for the 21st Century."
154. "Ólafur Ragnar Grímsson: A New Model of Arctic Cooperation for the 21st Century."
155. Mia Bennett, "Arctic Circle 2022: A NATO admiral, Chinese diplomat, and Faroese metal band walk into a concert hall," *Cryptopolitics* (blog), October 19, 2022, accessed January 18, 2023, https://www.cryopolitics.com/2022/10/19/arctic-circle-2022/. Ólafur Elíasson is a world-famous Icelandic-Danish artist.
156. See, for example, "Harpa er martröð skattgreiðenda" [Harpa Is the Nightmare of Taxpayers], RUV, August 4, 2012, accessed February 20, 2024, https://www.ruv.is/frettir/innlent/harpa-er-martrod-skattgreidenda.
157. Interview with a senior Russian official, November 3, 2017.
158. Interview with a senior Russian official, June 22, 2009.
159. See the Arctic Circle website, accessed January 10, 2023, https://www.arcticcircle.org/.
160. "Forsetinn setti ofan í við norskan ráðherra" [The Icelandic President Scolded a Norwegian Minister], *Vísir*, March 19, 2014, accessed January 9, 2022, https://www.visir.is/g/2014140318637/forsetinn-setti-ofan-i-vid-adstodarutanrikisradherra-noregs.
161. "Þurfa að skoða nýjar aðferðir í samskiptum við Rússa" [New Approaches in the Relationship with Russia Need to Be Examined], *RUV*, March 29, 2022, accessed January 9, 2024, https://www.ruv.is/frett/2022/03/20/thurfa-ad-skoda-nyjar-adferdir-i-samskiptum-vid-russa.
162. "Ólafur Ragnar deilir viðtölum við Rússa um stríðið" [Ólafur Ragnar Grímsson Circulates Interviews with Russians about the War], *Fréttablaðið*, April 5, 2022, accessed January 9, 2024, https://www.frettabladid.is/frettir/olafur-ragnar-gagnryndur-fyrir-ad-deila-skodunum-russa-um-stridid/.
163. "Ræða utanríkisráðherra [Þórdís Kolbrún Reykfjörð Gylfadóttir] á Helsinki Security Forum á þingi Hringborðs norðurslóða" [Address by the Icelandic Foreign Minister Þórdís Kolbrún Reykfjörð Gylfadóttir at the Arctic Circle's Helsinki Security Forum], October 17, 2022, accessed January 12, 2023, https://www.stjornarradid.is/raduneyti/utanrikisraduneytid/utanrikisradherra/stok-raeda-utanrikisradherra/2022/10/17/Raeda-utanrikisradherra-a-Hringbordi-nordursloda-Arctic-Circle/.

164. See "Statement of the Government of the People's Republic of China and the Government of the Kingdom of Norway on Normalization of Bilateral Relations," December 16, 2016, accessed January 10, 2023, https://www.regjeringen.no/globalassets/departementene/ud/vedlegg/statement_kina.pdf.
165. See Malgorzata Smieszek, Timo Koivurova, and Egill Thor Nielsson, "China and Arctic Science," in *China and Arctic Science*, ed. Timo Koivurova and Sanna Kopra (Leiden: Brill, 2020), 42–61.
166. He was succeeded as CEO by Ásdís Eva Ólafsdóttir.
167. Other board members are Sveinbjörnsson, Frederik Paulsen, the polar explorer and non-executive chairman of Ferring Pharmaceuticals, Brynhildur Davíðsdóttir, Professor of Environment and Natural Resources, and Þorsteinn Þorsteinsson, Researcher in Glaciology at the Icelandic Meteorological Office.
168. See "UAE's COP28 president will keep role as head of national oil company," *Guardian*, January 13, 2023, accessed January 14, 2023, https://www.theguardian.com/environment/2023/jan/13/uae-cop28-president-sultan-al-jaber-to-keep-role-as-head-of-national-oil-company.
169. "Gagnrýnin byggir á vanþekkingu" [The Criticism is Based on Ignorance], *Morgunblaðið*, January 24, 2023. https://www.mbl.is/mogginn/bladid/grein/1827832/?t=838151989&_t=1674731166.3241313.
170. Other advisory board members include, among others, Lassi Heininen, Professor Emeritus of Political Science at the University of Lapland, Dana Eidsness, Director, Main North Atlantic Development Office; Mead Treadwell, the former Lieutenant Governor of Alaska; Milind Deora, former Union Minister of State, India; and Paul Holthus, President and CEO of the World Ocean Council; see Arctic Circle website, accessed March 2, 2023, https://www.arcticcircle.org/.
171. Ríkisendurskoðun [The Icelandic National Audit Office]. Útdrættir ársreikninga sjálfseignarstofnana og sjóða [Summaries of Annual Reports] fyrir Hringborð norðurslóða—Arctic Circle Foundation, 2013–2022.
172. Interview with a high-ranking Arctic official, February 6, 2024.
173. Summaries of Annual Reports for the Arctic Circle Foundation, 2013–2022.
174. Summaries of Annual Reports for the Arctic Circle Foundation, 2013–2022.
175. See, for example, Steinveg, "Governance by conference?" 247–251.
176. Steinveg, "Governance by conference?" 148.

CHAPTER 4

Geopolitical Divisions Over Ukraine: The Impact on Iceland's Arctic Policy

Some scholars[1] have defined contemporary Arctic geopolitics as an uneasy interaction between liberal institutionalism, emphasizing interstate cooperation and its ability to mitigate conflict, on the one hand, and neo-realism, with its preoccupation with the "anarchic" international system, power politics, and national security interests, on the other.[2] Thus, the reopening of the Arctic from the military confines of the Cold War was interpreted during the last decade of the twentieth century as an opportunity to establish a new cooperative political order. In the twenty-first century, however, the Arctic was sometimes defined in terms of a resurgent neo-realism as states allegedly scrambled to reterritorialize a space to pursue national goals as part of resource competition. A case can be made for such a dualist reading, but with qualifications. The immediate post-Cold War period witnessed, as we have seen, a period of collaboration in the Arctic. Subsequently, the spate of media accounts on the "Scramble for the Arctic" and the potential for Great Power rivalry temporarily disturbed this narrative. It was, however, quickly dwarfed by a revival of government and scholarly discourses on neo-liberal cooperation schemes underscored by the commitment of the Arctic Five to the Law of the Sea and a peaceful settlement of potential territorial disputes.[3]

The tension between the West and Russia over the latter's annexation of Crimea tilted the balance toward the conflictual, even if it did not lead to a rupture in the Arctic. The imposition of Western economic sanctions

© The Author(s), under exclusive license to Springer Nature Switzerland AG 2024

V. Ingimundarson, *Iceland's Arctic Policies and Shifting Geopolitics*,
https://doi.org/10.1007/978-3-031-40761-1_4

on Russia resulted in Russian retaliatory measures, including a refusal to approve the European Commission's observership in the Arctic Council, despite its fulfillment of the precondition expressed at the 2013 Kiruna ministerial meeting—namely, to come to an agreement with Canada over the resumption of seal exports by the Inuit people.[4] The worsening relations between the West and Russia raised questions about the narrative of Arctic exceptionalism without challenging its hegemonial status, for the military effects could also be felt in the North. Having ignored the North Atlantic in its 2011 maritime strategy, NATO began, in 2016, to pay, again, attention to the region. It followed increasingly vocal calls by Western military think tanks and commentators in the wake of the Ukrainian crisis to respond to Russia's military activities. Nostalgic memories of the Alliance's "golden age" were evoked by referring to its founding mission to convey a sense of common purpose against a well-defined adversary.[5] The nebulous post-Cold War reading of Russia as "neither friend nor foe"—to use Carl Schmitt's dichotomy[6]—fell out of favor, with Russia now being increasingly projected as a "revisionist" power intent on breaking the "rule-based" order.[7]

The argument was made that the bolstering of Russian maritime power—in addition to the refurbishing of former Soviet bases in the Arctic circle—could pose risks to Western security interests.[8] Yet a more assertive discursive slant about a potential Manichean maritime contest in the North had more to do with a Cold War past than with a post-Crimean geopolitical reality. It ruled out, for example, the possibility that Russia might be behaving differently toward bordering states—with which it shared historical trajectories and cultural affinities—than toward other states and regions. While Russia could be described as being a "revisionist" power with respect to Ukraine given its violation of the latter's internationally recognized borders, it still behaved more like a "status quo" one in the Arctic, maintaining its support for the Arctic Council and UNCLOS as underpinning the legitimacy of the Arctic governing framework.[9]

In an effort to maintain regional stability, the Arctic Eight, in fact, continued to promote intergovernmental collaboration despite the spill-over effects of the Ukraine crisis. Yet, instead of resorting to a normalization discourse—through clichéd metaphors of "peaceful Arctic family relations" or, conversely, of a "friend/foe" binary—they opted for some distance and ambiguity when describing Arctic conditions. There was no breakdown in communications between Russia and the Western Arctic states. Russia did not withdraw from multilateral Arctic political structures or refrain from abiding by mutually accepted regional norms. True, military collaboration between NATO states and Russia was suspended, with the latter withdrawing from the Arctic Security Forces Roundtable, which had been established in 2010 to promote collaboration among militaries that operate in the Arctic. The two sides continued, however, to work together in the field of civil security, maritime safety, and search and rescue in the Arctic.

This calculated blandness was reflected in Iceland's Arctic policies where a language of condemnation was balanced with that of continuity. The Center-Right government that succeeded the Left-Wing government in 2013—under Prime Minister Sigmundur Davíð Gunnlaugsson—did not introduce any new Arctic initiatives, preferring, instead, to stick to those put forward by its predecessor and to the 2011 Arctic policy. At the same time, the Ukrainian crisis resulted in a hardening of Iceland's political stance toward Russia. Following the lead of Iceland's NATO and Nordic partners, Foreign Minister Gunnar Bragi Sveinsson was an early critic of Russia's role in engineering the Crimean secession by force and of its direct involvement in the civil war in Eastern Ukraine; he also committed the Icelandic government to participation in the Western sanction regime against it.

Russia initially opted not to retaliate against Iceland by exempting it—and some other non-EU countries—from its trade ban.[10] In the preceding years, a sizeable rise in fish exports, mainly mackerel, to Russia had made it a most valued trading partner, with 10% of Iceland's total trade. It was, therefore, an economic blow when the Russians reversed course and extended their trade ban to all those countries that were involved in the anti-Russian punitive measures. It raised questions both within and outside the Icelandic government about whether Iceland should have been part of the Western sanctions imposed on Russia in the first place.[11] Under pressure from the powerful Icelandic fishing industry, Prime Minister Gunnlaugsson and Minister of Finance Bjarni Benediktsson, the head of the Independence Party, suggested publicly that Iceland's position could change. Gunnlaugsson even promised to try to get a better deal from the EU on tariff reductions to compensate for the loss of the Russia trade. They argued that through Iceland's participation in an ineffectual sanction regime, it was paying a steeper price than many other larger European countries that had shielded their own key industries.[12] As it turned out, these were empty words. In the absence of any vocal domestic political opposition to Iceland's participation in economic sanctions against Russia, there was no chance that Iceland would break ranks with its European and U.S. partners or that it would soften its inculpatory rhetoric about Russian actions in Ukraine. Sveinsson asserted that Russia's territorial grab in Ukraine was a violation of international law, which could not be condoned irrespective of the economic damage.[13] He, then, joined the defense ministers of the other Nordic states to pledge increased defense and security cooperation in response to Russia's Crimea annexation, which was justified on the grounds that the security situation in the Nordic countries' "adjacent areas"—as it was termed—had significantly worsened.[14]

Yet, while accepting the need for increased Western military activities in the North Atlantic, the Icelandic government did not, initially, tie them directly to the Arctic.[15] What was more, no efforts were made to abandon Iceland's emphasis on keeping the Arctic a peaceful area.[16] It is true that Lilja Alfreðsdóttir—who, in the spring of 2016, replaced Gunnar Bragi Sveinsson as Minister of Foreign Affairs after the Panama Papers scandal[17] had forced the resignation of Prime Minister Gunnlaugsson—warned in an address on the occasion of the ten-year anniversary of the U.S. military departure from Iceland against a "more assertive and unpredictable Russia undermining the international rule-based order."[18] Yet there was no interest in abandoning the notion of Arctic cooperation. The exceptionalist regional focus—where the Arctic was portrayed as a quasi-closed system—minimized the influence of external geopolitical developments and military threat scenarios. Thus, in the Icelandic case, the emphasis continued to be on civilian risk assessments, not military threats. The enormous search and rescue region of Iceland, which is 19 times the size of the country itself, continued to preoccupy the exclusively civil maritime preparedness system, especially the Coast Guard, its main component. The Coast Guard was highly dependent on regional and international collaboration with neighboring countries and within multilateral forums.[19] It was mostly concerned with improving its operational capability—in event of SAR and pollution prevention actions—in a vast and hazardous Arctic terrain. Despite Iceland's non-armed status, it had also become part of Arctic Security Forces Roundtable[20] as well as the Arctic Coast Guard Forum (ACGF) which was launched in 2015, with a focus on maritime security and safety cooperation among the Arctic states, including Russia, in the region.[21]

When the parliamentary committee responsible for developing Iceland's first cross-political National Security Policy—an initiative that had been launched by Skarphéðinsson when he was Minister for Foreign Affairs—finished its work, it did not want to abandon a broad understanding of the security concept. The policy, which was approved in the spring of 2016, defined environmental threats, sea pollution, and accidents due to increased maritime traffic in the Arctic as core risks facing the country.[22] As for hard security, it reaffirmed Iceland's commitment to NATO and the U.S.-Icelandic Defense Agreement. Iceland was, at this time, taking on an added importance in U.S. and NATO maritime surveillance activities in the North Atlantic and the Arctic. In the summer of 2016, a joint Icelandic-U.S. declaration, which was based on the 1951 Defense Agreement, was issued, permitting the United States to use facilities at its former Keflavík base and paving the way for increased rotational deployments of American troops to take part in anti-submarine warfare operations.[23] It was followed up by a bilateral agreement in late 2017, which provided for "access to airfields, ports, and other places within Iceland" as well as to "operating locations," including specific buildings, which had been returned to Iceland when the United States closed down the base.[24] These measures signified a new phase in the relationship with the United States and a return to a far more habitual or fixed military presence, but they did not elicit much domestic political attention or lead to a security policy debate in Iceland. In contrast to the Trump Administration, the Icelandic government did not even bother to publish the agreement, which suggests that it did not want to ruffle feathers.

What caused this U.S. shift in attitude toward the military importance of Iceland was rooted in the Ukrainian conflict. Starting in 2014, the U.S. military spent 20 days operating from Iceland to hunt for submarines; in 2017, the number of days had jumped to 153.[25] What this renewed militarization entailed was that it was becoming far more difficult to make a sharp political-geographic distinction between the North Atlantic as an area of military competition between the West and Russia and the Arctic as a separate "conflict-free" space. The case for such a division, which prior to the Ukrainian crisis had been articulated especially by Norway, allowed for resisting Russia in some areas and engaging with it in others, such as the Arctic.[26] It corresponded to Norway's earlier Cold War balancing between "deterrence" and "reassurance" in its policy toward Russia. In a joint newspaper article in March 2017, Guðlaugur Þór Þórðarson—who had become Iceland's Foreign Minister in January as part of the formation of a new Center-Right coalition government between the Independence Party and the liberal Reconstruction Party—and Ine

Eriksen Søreide, the Norwegian Defense Minister, only mentioned the North Atlantic, when reiterating the resolve of both states to collaborate more extensively in the field of defense. In a speech at a conference in Arkhangelsk, attended by Russian President Vladimir Putin, Þórðarson stuck to the script of portraying the Arctic region as having the potential to be a "venue for the Arctic states to set an example of how responsible actors on the international stage interact."[27]

Yet this regional demarcation obscured the military activities that were taking place in the Arctic as well as the North Atlantic; after all, the "GIUK gap" directly connects the two regions. It was further underscored when the U.S. Second Fleet was resurrected in 2018 seven years after it had been disbanded, for its task was not only to counter Russian naval forces in the North Atlantic but also in the Arctic.[28] This meant conducting area surveillance, including the deployment of so-called long-dwell drones; the deploying of maritime patrol aircraft, operating out of Iceland, to track Russian submarines; and running undersea monitoring systems or listening posts on the seabed.[29] The 2018 NATO exercise Trident Juncture, which had an Icelandic component, involved more than 50,000 troops in a "collective defense" exercise, extending to Arctic areas. The result was that the broad-based non-military aspects of Iceland's Arctic policy were increasingly being subordinated to a traditional military agenda.[30] Thus, while Þórðarson saw fit to praise, in a speech in 2018, the soft security cooperation among the Arctic Council states, he also raised "major concerns" about Russia's military ambitions in the Arctic.[31]

4.1 Iceland as an Arctic Subtext in the U.S.-China Rivalry

Having moved away from the War on Terror as the primary focus of its national security strategy, the United States renewed its attention to Great Power rivalry or what it termed "strategic competition" in 2018. As its National Defense Strategy made clear, every global domain was contested—whether air, land, sea, space, and cyberspace—including the Arctic.[32] The main state-based rivals singled out in this scenario were Russia and China, with the former engaging in a geopolitical and military contest with the United States and the latter striving for its future displacement as the preeminent world power. U.S. interest in the North must be seen within the context of this broader geopolitical outlook. As it had done during the Cold War, the United States began to take active measures to counter Russian submarine capabilities to exploit strategic routes from the Arctic into the North Atlantic.[33] China's presence in the Arctic was added to this antagonistic picture on the account of its interest in investing in natural resource exploitation and commercial and infrastructure initiatives abroad—notably, the Belt and Road Initiative (BRI) and its spin-off, the Polar Silk Road—which were seen as instruments to project political and economic power globally. U.S. Arctic policy was, at that time, less about the existing Arctic than its future place in a rivalry between the United States and China. Thus, it put China under the same rubric as Russia, even if the respective positions of the two adversaries were radically different in the region. Irrespective of its military competition with the West, Russia, with one-fifth of its landmass being north of the Atlantic Circle, was by far the largest Arctic state.[34]

While China had identified itself—rather presumptuously—as a "near-Arctic state" in its 2018 Arctic policy,[35] its geopolitical and economic interests could not be compared to those of Russia in the Arctic where it was militarily absent. What partly motivated the United States was to contain Chinese economic influence as a preventive strategy. At the same time, it flirted with the geopolitical option of establishing a Cold War-style "linkage"—that is, to tie China's behavior in the Arctic to that in other parts of the world, such as the South China Sea.[36] The idea had even been floated in a U.S. congressional report to "suspend" China's status as an Arctic Council observer state "as a punitive cost-imposing measure" if it engaged in "unwanted" actions in the South China Sea.[37] As has been mentioned, China has, indeed, made sweeping and controversial sovereignty claims in the South China Sea, which have been contested by other Asian states and undercut by a ruling against it by the Permanent Court of Arbitration (PCA) at The Hague.[38] Yet the United States was, again, also open to the criticism that it artificially transposed military threats in the South China Sea to the Arctic.[39] Moreover, U.S. insistence that the freedom of military vessels is a universally established practice embedded in international law—that states do not have the right to limit navigation control in their EEZs for security purposes—is not universally shared, with countries such as Brazil, Indonesia, India, and Argentina opposing it.[40]

In 2019, this U.S. Arctic narrative against Russia and China was reinforced in stark rhetorical terms by Secretary of State Mike Pompeo at the Arctic Council ministerial meeting in Rovaniemi, Finland.[41] Stemming from the lingering geopolitical effects of Russia's actions in Ukraine and the worsening of U.S.-Sino political and economic relations, the hardening of U.S. position influenced other states, including the Nordic ones. Thus, Norway and Denmark started investing more heavily in military capabilities for maritime purposes to counter Russia.[42] The Danes committed $250 million to the operation of two long-endurance drones from Greenland and to the reestablishment of Cold War-style early warning radar stations in the Faroe Islands. It was also widely reported that the Danish government's decision, in 2019, to finance airports that China had offered to build in Greenland was taken after a U.S. political intervention. As part of its rivalry with China,[43] Greenland had by that time regained its geostrategic value for the United States, which has run the Thule military base there since the early 1950s. The Norwegians also strengthened their naval presence in the North and increased their stocks of ammunition, fuel, and equipment. And U.S. marines were, temporarily, deployed to Northern Norway in a symbolic show of American interest and to bolster NATO's interoperability and military training in the region.[44]

As for Iceland, the importance of its strategic location had been underscored, when the United States resumed anti-submarine missions from the country.[45] In addition, the Americans began refurbishing military installations—which had been left behind when the Keflavík base was closed—and constructing new ones for the Navy and Air Force. The revival of U.S. military interest in the island was highlighted by high-level political symbolism. When Pompeo visited Iceland in 2019, he was the first U.S. Secretary of State to do so since Condoleezza Rice made a stop-over 11 years earlier.[46] Vice President, Mike Pence, also came to Iceland in 2019 over half a century after Vice President Lyndon B. Johnson visited the country in 1963.[47] Both were keen on warning Icelandic government ministers against the influence of China in the Arctic region, with Pence singling out Chinese multinational technological corporations. He even congratulated the Icelandic government—a Left-to-Right coalition government formed in late 2017 and made up of the Left Green Movement the Independence Party, and the Progressive Party—for refusing to take part in the Polar Silk Road project. Since it had made no such decision, he was quickly corrected, publicly, by Prime Minister Katrín Jakobsdóttir, the leader of the Left Greens, and conservative Foreign Minister Þórðarson after their talks.[48] Yet, to drive home the point that Iceland was back on the U.S. military map, three U.S. B-2 stealth bombers, capable of delivering both conventional and nuclear munitions, were for the first time deployed, in late summer 2021, to Iceland, where they operated for a couple of weeks.[49] In the eyes of the United States, Iceland and Greenland were potential "weak links" in a new geopolitical landscape. To reduce Chinese influence, the Pentagon announced its intention of pursuing "strategic investments" in the Arctic, including those that could serve dual military and civilian purposes, such as airport infrastructure projects in Greenland.[50] Since 2018, the United States has invested tens of millions of dollars in military infrastructure in Iceland as part of its expanded presence, even if Russia, not China, is the main target there.

The securitization of China's presence in the Arctic posed a dilemma for Iceland because of its erstwhile hedging strategy. The Icelandic government put, as we have seen, much emphasis on expanding its relations with China following the financial crash,[51] with Icelandic President Grímsson and Foreign Minister Skarphéðinsson leading such efforts. To be sure, after Iceland's economic recovery, successive Icelandic governments felt less material need to follow up on bilateral projects, preferring to stick to Iceland's traditional ties with its Western partners. During the second Obama Administration, the Americans also began to show more interest in cultivating the Nordic countries, inviting the five Prime Ministers to the White House, where they discussed security cooperation and continued sanctions against Russia after the Crimean annexation as well as an Arctic conservationist environmental agenda.[52] Another factor was the hardening and assertiveness of China's global posture. Western criticism of China's human rights record under its leader Xi Jinping, especially its hardline stance on protestors in Hong Kong and its repression of the Uyghurs in Xinjiang, dampened enthusiasm for deeper cooperation.[53]

The anti-Sino course adopted by the Trump Administration resulted in increased scrutiny by the United States and some other Western countries of bilateral initiatives between Iceland and China. Questions were, for example, raised about the construction of a China-Iceland Aurora Observatory, whose name was changed to Arctic Science Observatory (CIAO), at Kárhóll in the northeastern part of Iceland—focusing on solar-terrestrial interaction and space weather, climatology, glaciology—as well as Chinese investments in the Icelandic economy. Given that the CIAO's original purpose was limited to aurora research, its expanded functions evoked suspicions that research on satellite remote sensing could be used by China for military purposes. Having been initiated by Iceland, not China, this research project was part of a scientific agreement concluded between the two countries in 2012. The name change of the observatory was, in fact, suggested by the Icelandic partners who wanted to broaden the scope of the project because there were so few academics working on aurora research at the University of Iceland.[54]

After repeated delays, the construction of the research facility is still not finished. The Chinese have agreed to provide the $1 million needed to complete it after having experienced repeated cost overruns.[55] COVID-19 was an additional delaying factor—with no Chinese representatives visiting the Observatory from 2019 until the end of 2022.[56] Still, the CIAO was formally opened in 2018. On the occasion, former President Grímsson was seemingly concerned with ending the political speculations about the project. He stated that building the facility "in a place that is as open and democratic and transparent as Iceland, is a message in itself" from China about its commitment to scientific cooperation.[57] Situated not far from Akureyri and Húsavík, another town in the northeast of Iceland, the CIAO facility is owned by a local non-profit organization, the Arctic Observatory, which provides operational services for the center.[58] CIAO's research partners are Icelandic and Chinese institutions, universities, and non-profit organizations.

The building, which has cost close to $6 million, was funded by China through the PRIC without any counterpart financing by the participating Icelandic institutions despite their own instrumental role in initiating the project.[59] Since the Chinese were not allowed to buy land for the facility, the organization running it concluded, in 2018, a 99-year lease agreement with the PRIC, which was approved by the Icelandic Ministry of Justice. The length of the agreement—while still common as a matter of business practice, if no longer as the longest lease of a real property in many legal jurisdictions—attracted no attention when it was made. The two scientific research partners, PRIC and the University of Iceland's Science Institute, have joint access to all the scientific data—extracted from auroral all-sky imagers, producing images of the auroras as well as single-color images (spectral lines), an imaging riometer that measures the upper atmosphere opacity, and a vector-magnetometer that measures the three components of the local geomagnetic field. Much of the data is shared with other foreign and domestic academic institutions.[60] The facility has, however, not been fully operational. In the last few years, the Icelandic Foreign Ministry has stepped up its engagement with the CIAO project—a move that the Chinese welcomed because of the dispute with the Icelandic local operating organization over the facility's excess construction costs. Yet what complicated the situation was that the Ministry of Higher Education, Industry, and Innovation refused to get involved in the project because of its political implications. Hence, there was a reluctance within the Icelandic government to take responsibility for it.[61]

Western governments have questioned Chinese motives.[62] The United States claimed, for example, that such scientific facilities could facilitate dual-use research with intelligence and military applications in the Arctic.[63] Former Foreign Minister Þórðarson has revealed publicly that during his tenure from 2016 to 2021, he had expressed concerns about the Observatory on the grounds that it seemed to have been built and operated without monitoring by the government. Yet, as a minister, he does not seem to have used his position to influence or correct the situation. Indeed, a 2021 report sponsored by the Foreign Ministry under Þórðarson on economic opportunities in the Arctic—which was headed by a former Independence Party politician—was ironically entitled "Northern Lights," with a photo of the Arctic Observatory figuring prominently in it.[64]

Some Icelandic officials thought that the allegations by the Americans and others about Chinese nefarious intentions were overblown and slightly paranoid. They have pointed out that the U.S. representatives and those of other Western embassies in Iceland have been granted access to the facility on numerous occasions and that all science data collected at Kárhóll are fully accessible to Icelandic scientists.[65] This did not prevent two members of a Congressional Comittee on the Chinese Communist Party from reiterating their concerns about Kárhóll in a letter to the Secretaries of State and Defense in October 2024.[66] Apart from questions from Western partners, the government has been pressured on the China connection by an Icelandic newspaper, which is known for investigative journalism[67] and by an opposition parliamentarian who has also been critical of increased U.S. military activities in Iceland.[68] Former conservative Justice Minister Björn Bjarnason, the author of the Nordic security report, has gone further, calling, in a rather alarmist way, on the government to "erase the China stigma" either by cancelling the Free Trade Agreement or close down the Kárhóll research facility.[69]

Foreign Ministry officials have sought to depoliticize the issue by suggesting opening up the CIAO project to other partners through data sharing and research cooperation. Japanese, British, and Canadian scientists have, for example, conducted solar-terrestrial research in Iceland because it is considered a prime Northern location for such scientific investigations without any government oversight. There have been discussions about including them in the Observatory research activities, especially the Japanese who agreed to move their scientific aurora equipment to Kárhóll and conduct their research there.[70] The Canadians, who were also approached by the Icelanders, have suggested that any possible Canadian involvement would be contingent on a bilateral agreement with Iceland, where the Chinese would have no role.[71] The history of the project shows how the space for small-state hedging has shrunk due to the U.S.-China contest. What started out as a bilateral Chinese-funded science program in Iceland, which was also pursued by local interests in the Akureyri area as a northern lights business project, became mired in geopolitical calculations.

As for Chinese investments in Iceland,[72] they have, as discussed here, been few compared to other Western Arctic states, with no reported Foreign Direct Investment (FDI) in any of the three restricted sector-based areas: fishing, energy, and transportation. At first, the Icelandic government publicly refrained from parroting the Trump Administration's rhetoric against China and its presence in the Arctic. It stressed the value of increased trade with China, which Foreign Minister Þórðarson termed an "untapped potential, which we are eager to exploit."[73] Moreover, while the U.S. securitization campaign against the Chinese technology company Huawei was felt in Iceland—like in other European countries—it did not lead to a total exclusion from Iceland's digital infrastructure. Unlike Denmark and Sweden, which have effectively banned Huawei, Iceland, like Finland and Norway, has not gone so far. Still, Þórðarson raised "national security" concerns with respect to Chinese economic interests in Iceland—a position that was incorporated into the explanatory memorandum accompanying Iceland's revised Arctic policy in 2021.[74] The adoption of new investment screening mechanisms targeted at China in the other Nordic countries is also an indication of how they have distanced themselves from the Chinese in line with new geopolitical realities.

Another sensitive case, involving Iceland, is the so-called Finnafjörður project or the construction of a harbor-industrial location for future transarctic Arctic shipping in the northeastern part of Iceland. While dormant, it could possibly attract Chinese investors in the future if the interest in the transpolar route can be maintained. Since there is a NATO radar station close to the site,[75] the Icelandic Foreign Ministry has viewed the project as having defense and military implications and wants to have a say in how the site for the proposed transshipment port would be defined and operated.[76] While it has used search-and-rescue needs for the Icelandic Coast Guard to justify government control of the area, military reasons were the underlying motive. The United States does not want the Chinese to get an economic foothold in Finnafjörður, where a military harbor could be constructed in the future. When Admiral Richard Burke, the Commander of the U.S. Navy in Europe and Africa, warmed up to such a port in the eastern part of the country—during a 2019 visit—Prime Minister Katrín Jakobsdóttir rejected it out of hand, adding that no U.S. military base would be constructed in the area.[77] Despite her refusal, the Americans continued to express interest in the idea as part of their search for an operating naval base in the North.[78]

The rotational nature of U.S. military deployments to Iceland was used, at that time, by the Icelandic government to deny that a policy reversal had taken place on the non-permanent stationing of troops on the island. Foreign Minister Þórðarson was, however, ideologically committed to a pro-U.S. and pro-NATO line and supported an expanded military presence. True, he had to consider the anti-militarist attitudes of the Left Greens, in particular, and the non-belligerent sentiments of an unarmed nation, in general. Thus, he stressed that "our regions, the North-Atlantic and the Arctic," should not "become an area for re-armament" and that "low tension" should prevail "in the High North."[79] Still, such utterances did not prevent him from offering thinly veiled backing for the region's securitization by referring to the need to "avoid a security vacuum and ensure full situational awareness" in the region.[80] In his annual report on foreign and international affairs in 2019, he rationalized for the first time, publicly, the upsurge in Western military surveillance activities from Iceland by pointing to Russia's military buildup in the North[81]—a theme he returned to in the 2020 and 2021 reports.[82] While couched in defensive terms, his words showed how the original aim of Iceland's Arctic policy to fight the militarization of the Arctic was being recast. Instead, he adopted a more confrontational vocabulary, which he justified by a changed international outlook. Partly due to U.S. pressure, Iceland was slowly but steadily realigning itself with a Western geopolitical agenda against Russia and, to a lesser extent, China.

4.2 Cultivating Greenland in the "New Arctic"

Against the backdrop of the spotlight on Greenland in Arctic geopolitics and regional politics as well as of Iceland's relationship with it, Foreign Minister Þórðarson decided in 2019 to appoint former Foreign Minister Skarphéðinsson to chair a committee entrusted with finding ways to expand political, economic, and social ties between the two countries. Within a broader West Nordic focus, Skarphéðinsson had been keen on cultivating Greenland as testified to by the 2011 Arctic policy, the fishery and health agreements in 2012, and the decision, in 2013, to establish an Icelandic consulate in Nuuk. The Greenlandic government had reciprocated by opening its third diplomatic representation abroad in Reykjavík in 2018 after Brussels in 1992 and Washington, D.C. in 2014. It saw opportunities in developing closer relations with Iceland in the fields of commerce and trade, civil aviation—building on the operations of direct flights between the two countries—and tourism, with two new airports in construction at the time.[83]

Skarphéðinsson's Greenland committee produced a lengthy report—dubbed *Greenland in the New Arctic*—in December 2020, containing no fewer than 99 ideas for expanding bilateral ties.[84] In addition to existing projects in the areas of fisheries, aviation, tourism, and the Arctic, it singled out additional ones in health and social care, trade, education, and resource utilization. The report—which was written in a crisp style honed by Skarphéðinsson as an articulate and colorful former politician—eschewed diplomatic niceties toward the Danes in favor of full support for Greenland's "firm path towards independence."[85] It proposed that a bilateral framework agreement be concluded between Greenland and Iceland to put some of its suggestions into practice,[86] such as negotiating a trade agreement that would favor the Greenlanders by reducing "the prices of daily necessities for Greenlandic consumers"; financing a leisure center in the eastern part of the country to help combat social problems such as sexual molestations of children and suicide rates of young people; establishing a search and rescue team made up of volunteers and based on the Icelandic SAR model; implementing a joint distance learning program between the Ilisimatusarfik/University of Greenland and the University of Akureyri; concluding a fishery agreement between the two countries focusing on shared migratory fish populations; and facilitating the setup of small hydroelectric power stations in Greenland to substitute for oil.

Apart from putting forward such wide-ranging ideas, the report made much of the growing geostrategic importance of both countries. Greenland was depicted as a "zone of contest" between the United States and China—and Iceland as a site of renewed American military interest after a hiatus created by the 2006 military withdrawal.[87] Skarphéðinsson's old idea of the transarctic route—with Greenland now being assigned a central geopolitical and economic role together with Iceland—was recycled in the report. It quoted Ólafur Ragnar Grímsson's comment that "the central area is precisely in the spotlight of the world because of security interests of the Great Powers, the proximity to new shipping lanes, the importance of Greenland, which, of course, is the epicenter of climate change."[88] And it relativized, in euphemistic language, President Trump's colonizing bid to buy Greenland. Even if Trump had expressed his interest in the territory in an "unusual manner," it stated that he had been the first U.S. President to understand the "new geopolitical importance of the Arctic," which will be "part of his legacy" irrespective of how history will judge him.[89]

While highly supportive of the Greenlandic cause, the report also offered politically insensitive paternalistic advice. Icelanders were to provide technical knowledge and experience to a Greenlandic society in need of economic and social development. To raise the profile of Greenlanders internationally, it was suggested that the Icelandic Foreign Ministry "in cooperation with the Greenlandic government" approach the Arctic Circle Assembly—that is, Grímsson—for the purpose of establishing a think tank that would specialize in Greenland and the Arctic.[90] Many Greenlandic politicians have been grateful to Grímsson, who was honored by the parliament with Greenland's meritorious service medal, for his political support and for giving them an elevated platform at the Arctic Circle Assembly. Yet it stretched credulity to include a proposal—in a report on Icelandic-Greenlandic cooperation—that urged the Icelandic government to see to it that an Arctic Center, in the name of the former President, be constructed in Reykjavík as a future home for the Arctic Circle Assembly and as an international research and exhibition facility. It raised the awkward question of whether the purpose was to turn an Icelandic personal legacy and nation branding project into a goodwill gesture toward the Greenlanders.[91]

Unsurprisingly, Danish officials were critical, if not publicly, of the Greenland report, especially those parts that referred to Denmark's colonial past or to the inevitability of Greenland's independence. In contrast, the reaction of the Icelandic government and parliament was highly favorable. Four months after the report's publication in early 2021, a parliamentary resolution—based on its findings—was passed unanimously.[92] The parliamentarians decided, however, that both governments should earmark bilateral priorities in Greenlandic-Icelandic cooperation and that the 99 proposals be seen as reference points rather than commitments.[93] Shortly before the approval of the resolution, the Icelandic government dutifully committed itself, formally, to the establishment of the Ólafur Ragnar Grímsson Center on the Arctic. Subsequently, Skarphéðinsson was appointed its first chair.

While the Greenlandic government responded positively to the Greenland report, it did so in cautious and general terms. Privately, the Greenlanders were not happy with its emphasis on social problems in Greenland, which Icelanders should help fix.[94] In a joint declaration, Foreign Minister Þórðarson and the Greenlandic Minister for Foreign Affairs and Climate Pele Broberg stated that the report provided "a solid foundation for identifying areas of increased and new cooperation."[95] That such a bilateral initiative was backed by the Greenlandic population was confirmed in the first foreign and security public opinion poll conducted in Greenland, which was published in 2021. According to it, 90% of Greenlanders favored increased cooperation with Iceland and only 4% less.[96] No other country scored better in the poll. When Prime Minister Katrín Jakobsdóttir visited Greenland in the spring of 2022—the first time in 25 years that an Icelandic Prime Minister had done so—Greenland's Prime Minister Múte B. Egede suggested that increased Icelandic-Greenlandic relations be formalized in some way.[97] In the fall, Egede and Jakobsdóttir issued a common "declaration of cooperation" between Greenland and Iceland. Instead of a specific framework agreement—as proposed by the Greenland committee and the Icelandic parliament—the declaration was presented as providing the "overall framework for our bilateral cooperation" in seven priority areas: trade, fisheries, economic cooperation, climate change and biodiversity, gender equality, education and research, and cultural cooperation.[98] This declaration now forms the basis for Iceland's relations with Greenland. While both sides expressed a desire to do the groundwork for a Free Trade Agreement, no new commitments have been made. In the bilateral declaration, it was pointedly stated that the feasibility of such trade and fishery agreements was to be explored based on "equality."[99] While the declaration was a start, it fell short of the ambitions spelled out in the Greenland report—a report that reflected a genuine interest in strengthening bilateral relations but was written through the cultural prism of myopic Icelandic perceptions, understandings, and aspirations.

4.3 Iceland's Chairing of the Arctic Council in Pandemic Times

When, in the spring of 2019, Iceland took over the chair of the Arctic Council from Finland, there were fears that there would be a political fallout from U.S. Secretary of State Mike Pompeo's hostile intervention at the Rovaniemi ministerial meeting. Such meetings had never been used before as a venue for geopolitical attacks. Pompeo had focused on China in his criticism, but Russia came close second, and he did not even spare a neighboring ally, Canada. He mocked China's self-characterization as a "near-Arctic state" by pointing out that the shortest distance between China and the Arctic comprised nine hundred miles.[100] There were "only Arctic states and non-Arctic states" because no "third-party category existed," and "claiming otherwise entitled China to exactly nothing."[101] Warning against "China's aggressive behavior," he stated that it could use its scientific research presence in the Arctic to strengthen its military presence, including the deployment of submarines.[102] While acknowledging that Russia cooperated with other Arctic states in a number of areas, such as conservation, he also attacked it for its truculent behavior and for reestablishing military bases in the Arctic. Finally, he did not only reiterate U.S. non-recognition of Russia's sovereign claim to the Northern Sea Route but also reminded Canada that the same applied to its claim to the Northwest Passage. By doing so, he broke the "silent pact" of not dwelling publicly on the long-standing U.S.-Canadian dispute over this route.[103]

Pompeo's outburst, it turned out, did not have an effect on Iceland's tenure as the chair of the Arctic Council. At the Rovaniemi meeting, where Þórðarson presented Iceland's program, he did not veer from his prepared text on the Arctic Council's importance as a venue for dialogue and "peaceful cooperation" and on the need to avoid a military buildup or conflicts in the North, even if he made it clear—in an obvious allusion to Russia—that increased military activities in the North Atlantic and the Arctic were a source of concern.[104] In contrast to Pompeo—who had refused to sign on collective goals for the reduction of black carbon, while, simultaneously, boasting about the U.S. environmental record—Þórðarson spoke of the detrimental consequences of climate change, especially for the Indigenous communities. As for Iceland's chairing priorities, the focus was—as was to be expected for a state that relied on fisheries—on the oceans, with references to pollution and ways to mitigate it as well as to marine resources, especially the need to increase the utilization level of biomass. Moreover, Iceland promised, in general terms, to work on improving Arctic governance by strengthening the inner workings of the Arctic Council, not least between the Arctic states and the permanent participants from Indigenous groups as well as with other external stakeholders in the Arctic. Climate issues and green energy transition solutions for Arctic communities were highlighted in the program. Sustainability was to be promoted through cooperation on "gender equality, connectivity, adaptation and resilience."[105] Such jumbled and generally worded aims—with different concepts being indiscriminately lumped together—did not offer much in terms of concrete proposals. But since the AC's decisions and statements require a consensus, the chairing country is rarely in a position to promote an ambitious agenda.

One thing the Icelanders wanted to do during their tenure was to establish closer ties between the Arctic Council and the Arctic Economic Council (AEC). At the 2013 Kiruna ministerial meeting, the Arctic states had decided to create a task force—made up of representatives of Russia, Canada, Finland, and Iceland—to explore the idea of a circumpolar business forum. On its recommendation, the Arctic Council established, in 2014, the Arctic Economic Council (AEC) as an independent forum to promote "sustainable" business ventures between Arctic states and the Indigenous communities in the Arctic. The AEC's executive board is headed by a representative from the state that holds the rotating chair of the Arctic Council, but it is a separate entity. Given the AEC's pro-business orientation, it is preoccupied with resource utilization, not preservation in the Arctic—and has resisted what one of its spokesmen termed "black/white" approaches toward economic development in the region. Thus, it has taken a dim view of attempts by banks to ban or reduce energy investments in the Arctic due to climate change.[106] The Icelandic businessman Heiðar Guðjónsson—who was the chair of the AEC when Iceland led the Arctic Council, and who headed the Icelandic company involved in the abandoned Dragon Area oil exploration initiative with the Chinese and Norwegians—put it this way: "In the international debate, the environment gets all the attention. But there are important cultures that have been persistent in the Arctic for thousands of years, and they deserve to get attention and opportunity to develop economically like the rest of the world."[107] Thus, the AEC's exploitation agenda was projected as reflecting local community interests and needs. Under the Icelandic Arctic Council chairmanship, the AC and the AEC held, in October 2019, their first joint meeting in Reykjavík, which centered on the respective agendas of the two bodies. Echoing Iceland's long-standing discourse on the need to "balance" environmental protection and economic use in the Arctic, Einar Gunnarsson, the Chair of the Arctic Council's Senior Arctic Officials, stressed, at the meeting, that such an approach would be of mutual benefit.[108]

As it turned out, the social restrictions adopted to respond to the COVID-19 pandemic overshadowed Iceland's Arctic Council agenda in 2020–2021.[109] The Icelanders singled out the fight against ocean pollution as one of the accomplishments of their chairing the AC, even if it was a modest one.[110] While Iceland's proposal for the Arctic Council Strategic Plan 2021–2030 was approved, it was, to a large extent, based on ideas that had been developed during the preceding Finnish AC chairmanship. The strategic plan is a cautiously worded document—with references to climate change, eco-systems, the marine environment, social and economic sustainability, knowledge production and communication, and a "stronger Arctic Council"—but without listing commitments to specific goals.[111] The Reykjavík Declaration of the AC ministerial meeting in May 2021 was, in contrast, a long text that in a self-gratulatory tone reaffirmed, in 62 numbered paragraphs, an idealistic common agenda by the Arctic Eight, which included the maintenance of peace, stability, and constructive cooperation in the Arctic, the promotion of social and economic developments, and a green environmental agenda aimed at safeguarding the marine environment and fighting ocean pollution.[112] What made the Reykjavík ministerial different from that of Rovaniemi was not what the Icelanders had achieved when chairing the Arctic Council but the attitudes of the Americans. The Biden Administration eschewed the Trump Administration's anti-climate change rhetoric and refrained from using the meeting to attack Russia or China.

When Russia took over the chairmanship of the Arctic Council at the Reykjavík ministerial meeting, Russian Foreign Minister Sergey Lavrov struck a conciliatory tone. He hailed the approval of the Arctic Council's strategic plan—"the first ever long-term planning document reflecting the member-states' shared vision of the goals to be dealt with in the coming decade"—and commended "all our partners" who shared Russia's view that the Arctic was a "territory of peace, stability and constructive cooperation."[113] Moreover, he argued that "the positive relations that we have within the Arctic Council" should be extended to the military sphere and that a multilateral dialogue between the Chiefs of General Staff of the Armed Forces of the Arctic States be resumed.[114] Russia had been suspended from this forum in 2014 in response to its annexation of Crimea. Antony Blinken, the new U.S. Secretary of State, had expressed a critical view of Russia, rejecting its proposal for new navigational rules in the Arctic on the grounds that they were inconsistent with the Law of the Sea and accusing it of a military buildup in the region. Thus, the Americans were not ready to resume military talks with the Russians. Blinken also reiterated U.S. opposition to the resumption of a military dialogue among all the Arctic states.[115] Yet, at their first person-in-person meeting since the Biden Administration took office in January 2021, Blinken and Lavrov pledged to work together despite their differences.[116] Thus, the constructive spirit of the Reykjavík ministerial stood in marked contrast to the one in Rovaniemi.

As for the Icelandic hosts, the ministerial was all about the symbolic politics of place. Referring to the Reagan-Gorbachev Summit, Þórðarson stated that the Icelandic government wanted to do its part to facilitate improved relations between the United States and Russia.[117] Ólafur Ragnar Grímsson boasted that Iceland had cemented its place as the "Arctic's central hub." To him, the Reykjavík Declaration would be seen as a "seminal factor" in Arctic developments, pointing to its allusion to the region as an example of peaceful cooperation.[118] In addition, he claimed that the AC's strategic plan represented an innovation not only within an Arctic context but, "in a certain sense," also a global one.[119] It was "almost unprecedented" that in times of conflict and dissolution states could agree on such a wide-ranging plan, especially when the United States and Russia were among them.[120] Even if Grímsson could not possibly have foreseen Russia's invasion of Ukraine less than a year later, his hyperbole hardly applied to the unambitious strategic plan. The Reykjavík ministerial meeting had, of course, nothing to do with Iceland as an Arctic hub; it was part of an Arctic Council's routine schedule and rotating chairmanship. Such political interventions mixed a small-state mentality with nation branding. Harking back to the Reagan-Gorbachev encounter—or even earlier to the 1972 Fischer-Spassky world chess championship with its Cold War symbolism[121]—the purpose was, yet again, to tie Iceland to events of global significance.

4.4 The Securitization of Iceland's Arctic Approach

In May 2021, the Icelandic parliament approved a revised cross-political Arctic policy.[122] As was the case with the original one in 2011, it stressed the role of the Arctic Council as the most important forum for the region—and to ensure peaceful resolution of disputes, it reiterated that international law, including UNCLOS as well as international human rights treaties, should be respected. Given the increased global focus on climate change, material factors figured less prominently in the new document than in the old. The policy pledged to make sustainable development a guiding principle based on the UN goals, stressed the importance of the 2015 Paris Climate Agreement, and called for the extension of the ban on the use of heavy fuel oil within Iceland's territorial waters to the Arctic as a whole. This did not mean, though, that economic opportunities in the Arctic should be shunned but rather pursued from an "ecological sustainability perspective" and through what it termed "responsible utilization of natural resources."[123] In line with the West Nordic focus of the 2011 policy, Greenland and the Faroe Islands were mentioned as partners in promoting trade, commerce, services, and education in the region. The international profile of the West Nordic Council had been raised after it was granted an observer status in the Arctic Council in 2017,[124] even if its efforts to forge a common West Nordic Arctic policy had not materialized. The policy also referred to the 99 proposals listed in the Greenland report. And it articulated the need for extending Nordic policy coordination on the Arctic, especially security cooperation between the Nordic states, alluding to the Björn Bjarnason report.

The non-military security component of the 2021 text highlighted the importance of the Arctic Council members' agreements on search and rescue and against oil spills. It also pointed to ocean pollution in the Arctic—whether involving oil, chemicals, plastic, or poisonous materials—and to Icelandic commitments to fighting climate change. Finally, the policy committed, formally, to building "upon the success" of the Arctic Circle project and to the creation of a future framework for it through the establishment of a non-profit foundation that would operate an Arctic center in Iceland.[125] It reflected the desire of the political elite to capitalize on the branding of Iceland as an international conference hub on the Arctic. Still, no attempt was made to address the tension between Grímsson's open Arctic global agenda and the Icelandic government's regional one. The policy made it clear that Iceland's positive view of "growing interest in matters concerning the Arctic region from parties outside the region" was dependent on their respect for "international law and the status of the eight Arctic states."[126]

When compared to the 2011 Arctic policy, there were three substantial departures in the 2021 one. First, there was no reference to Iceland's status-seeking demand to be counted as a coastal state. The five littoral states had not been willing to accept the demand since Iceland did not have any extended continental shelf claims in the Arctic Ocean. Yet some of them recognized that the location of Iceland, with its EEZ including an Arctic maritime area, made it difficult to exclude the country from deliberations on key Arctic issues. No explanation was given, publicly, on the part of the Icelandic Foreign Ministry or Parliament for the reason for dropping the coastal state claim. Yet it was seen as being unnecessary after the Arctic Five had invited Iceland—together with the European Union, China, Japan, and South Korea—to discuss initiatives under the so-called five-plus five format,[127] particularly the 2015 "Declaration concerning the prevention of unregulated high seas fishing in the central Arctic Ocean"—which resulted in an intergovernmental agreement in 2018.[128] The Americans, with Canadian support, had pushed the issue, at least partly because they feared that if unchecked, the Chinese distant water fleet—the largest in the world—which had been involved in "illegal, unreported and unregulated" (IUU) fishing could run havoc in the Arctic Ocean in the future.[129] Over two thousand scientists had also called for such an agreement.[130]

Iceland had originally reacted sharply to the 2015 Arctic Five declaration on the central Arctic Ocean, with Minister of Foreign Affairs, Gunnar Bragi Sveinsson, summoning the ambassadors of the United States, Denmark, Canada, Norway and Russian to inform them that Iceland was not bound by it.[131] It was not made in the name of any international institution and was, in the Icelandic government's view, not consistent with international law. Iceland also protested being excluded from the deliberations.[132] When the Arctic Five relented and decided to include Iceland in further negotiations on a central Arctic Ocean agreement, they ensured that the Icelanders would be part of this future-oriented conservation effort. In addition, they sometimes used the term "central Arctic Ocean states" to describe the status of the Arctic Five to placate the Icelanders[133] who had reinforced their coastal state claim during the negotiations.[134] By distinguishing between the central Arctic Ocean and the Arctic Ocean made it easier to classify Iceland as an Arctic coastal state for maritime issues outside of the central Arctic Ocean.[135] Yet, as Jon Rahbek-Clemmensen and Gry Thomasen have pointed out, an alternative explanation may have been as important for the Arctic Five: that Iceland's inclusion in the agreement could make it more effective in practice and increase the chance that it would be respected by non-Arctic Five states.[136]

The 16-year International Agreement to Prevent Unregulated Fishing in the High Seas of the Central Arctic Ocean, which was meant to enhance the legitimacy of a fishing moratorium before a regulatory framework was in place,[137] went into force in 2021. Iceland's participation in it did not signal a policy change with respect to its broader claims—through fishery agreements with other states or regional fisheries management organizations—to a share in fishing activities, even if fish stocks shifted between areas due to climate change or changed conditions in the marine environment. That was one of the Icelandic fishing industry's main concerns: that a moratorium could and would weaken its access to fishing grounds outside its EEZ. What is more, when Iceland ratified the agreement it reiterated its legal stance on the Svalbard dispute and on restrictions on Norwegian sovereignty over the archipelago due to the special rights accorded to the parties to the Spitsbergen Treaty.[138]

Second, Iceland's commitment to Indigenous rights was weaker in the 2021 policy document. In the 2011 parliamentary resolution, support was expressed for Indigenous participation in decision-making in all fields—whether political, economic, social, environmental, or cultural. While the revised Arctic policy backed the rights and equality of the Indigenous peoples and efforts "to protect the cultural heritage and languages of the nations that live in the Arctic," it was more concerned with Arctic inhabitants, in general.[139] In the explanatory note, it was stated that Iceland had—in its dealings with other Arctic states—supported the demands of the Indigenous peoples to be part of "political and economic" decision-making without specifically mentioning social, environmental, or cultural rights.[140] On the surface, this "diplomatic" line fell short of the open and unequivocal support voiced—in the Greenland report—for the independence aspirations of the Greenlanders. What complicates the matter is that the Greenlanders themselves have been putting more emphasis on self-determination rights guaranteed by the Act of Self-Government—giving them the option to pursue a path of independence—than on Indigenous or decolonization rights.[141] Nonetheless, the general de-emphasis on Indigeneity in the 2021 policy document gave the impression that the main focus should be on the four million people who lived in the Arctic rather than on the four hundred thousand Indigenous inhabitants.

Third, the elevation of potential Russian and Chinese threats confirmed the end of Iceland's non-Western hedging efforts harking back to the U.S. military withdrawal and the financial crisis. True, the 2021 policy reiterated that Iceland—like the other Arctic states—sought to keep the Arctic a low-tension area, where peaceful cooperation and respect for international law and sovereignty interests were maintained. Nonetheless, in the explanatory note, increased regional tensions were attributed to

Russia's military buildup and the West's response to it. Iceland itself had been part of increased militarization of the Arctic by allowing U.S. anti-submarine planes to operate from the country. While conceding that Russia had legitimate security and defense concerns in the Arctic, it was stated that the strengthening of its military capabilities went "considerably beyond what is warranted by circumstances."[142] Hence, the need for a NATO role in the North Atlantic as well as in the Arctic.

The policy memorandum also singled out China's interest in pursuing economic opportunities offered by a more accessible Arctic, not only with respect to natural resource extraction but also to the opening of a new sea route between China and Europe. But apart from being spurred by economic and science motives, China's involvement in the region was described as having military or "security and political dimensions, which needed to be given special consideration."[143] Thus, the Arctic policy mixed the old Arctic cooperation narrative with a competitive one, where a lightly drawn, if unmistakable, adversarial picture was drawn up of Russia and a potential one of China. It shows that breaks between periods do not necessarily involve complete changes of policy content but rather reconfigurations of pre-existing elements. The institutionalist, exceptionalist, and depoliticized characteristics that were still present in Iceland's policy toward the Arctic were steadily weakening. The trend toward securitization was a harbinger of the paradigm shift that took place after Russia's invasion of Ukraine.

4.5 Arctic Uncertainties: The Russian Invasion of Ukraine

When Iceland's coalition government parties—spanning the Left-Right spectrum—decided to renew their cooperation in late November 2021 after retaining their majority in parliamentary elections, they did not set any new policy accents on the Arctic. The government platform stressed—ritualistically—the need for strengthening Iceland's position as an Arctic state and for raising domestic expertise on Arctic issues, especially in the fields of education and research.[144] This idea of normality was abruptly upset three months later, when Russia invaded Ukraine on February 24, 2022. Declaring full solidarity with the Ukrainians, the Icelandic government immediately committed to the Western sanction regime against Russia, and it became involved in transporting weapons from NATO countries to Ukraine.[145]

Consistent with this stance, Iceland also joined the other Western members of the Arctic Council in suspending cooperation with Russia. When the new conservative Foreign Minister Þórdís Kolbrún Reykfjörð Gylfadóttir addressed the Arctic Circle Assembly in October 2022, she stated that "Russia's brutal war on a democratic, sovereign state" had made any cooperation with it in the Arctic Council impossible.[146] Portraying the Ukraine War in ideological terms as a contest between democratic and authoritarian forces, she added that the "democratic" Arctic Council members "must continue their cooperation to the extent possible." In response to "Russia's military build-up in the Arctic" over the "past decade and a half" and its growing military capabilities, she added that the West had responded with NATO's new Strategic Concept, new Arctic strategies, and increased military budgets.[147] She even used the term "Keflavik Air base" to describe Allied military "posturing and activities" in Iceland, even if the base had not been, formally, resurrected after its closure.[148] Finally, in an unmistakable swipe at China, she remarked that "we also need to be clear eyed about strategic interest of third parties that seek foothold in the North."[149]

What was suggested—in diplomatic terms—in the 2021 Arctic policy, thus, morphed into a hardline discourse following Russia's invasion of Ukraine. It also contrasted sharply with Grímsson's inclusive Arctic Circle ideology or his soft touch vis-à-vis Russia and China. From now on, Iceland's foreign policy, which had been gradually turning back to a traditional transatlantic orientation, became more heavily focused on a Western military posture against Russia. While left-wing Prime Minister Katrín Jakobsdóttir continued to plead for a peaceful Arctic, she expressed unreserved criticism of Russia's aggression against Ukraine. Gylfadóttir used the breakdown in the relationship between the West and Russia to reconfirm a firm pro-NATO line and to highlight increased U.S.-Icelandic defense cooperation in the North Atlantic and Arctic. P-8 bomber aircraft were, at that time, operating from Iceland all-year round. Yet she was careful about not overstepping government policy about a "non-permanent" foreign military presence in Iceland by emphasizing that no negotiations were taking place about the "return" of U.S. military forces.[150] Some Icelandic military commentators had been calling for the exploration of such an option[151] or even—less credulously for a country that has been without a military for centuries—for the establishment of an Icelandic army following Russia's invasion of Ukraine.[152] The Left Greens have, however, not formally abandoned their opposition to Iceland's NATO membership, even if they did so implicitly by voting in favor of an upgraded National Security Policy in early 2023, which reemphasized the importance of the Alliance and the U.S.-Icelandic Defense Agreement.[153] In 2016, the Left Green Movement had abstained in the voting on the cross-political National Security Policy.

Thus, the Ukraine War did not lead to formal declaratory changes in Iceland's foreign and security policy. Gylfadóttir even asserted that the 2006 U.S. decision to withdraw from Iceland had not been a "mistake"—as if she were approaching the question from an American perspective—adding that there were no plans for establishing an Icelandic military.[154] Her own party, the Independence Party, had, as has been elaborated on, fought tooth and nail against the U.S. plans to close the base. What was more, U.S. policy toward Iceland had undergone a fundamental change since the mid-2000s, with some American officials seeing the military departure, indeed, as a "mistake."[155] It soon also became clear that the U.S. military presence in Iceland would be increased further. In the spring of 2023, the Icelandic government authorized—for the first time—nuclear-powered U.S. submarines to have service visits and to replace crews off the coast of Iceland. The permission was conditional, for the submarines were not allowed to carry nuclear weapons in Icelandic territorial waters or to make port calls. The decision, which had been delayed for some time because of the political sensitivity of the issue within the government, was justified as a contribution "to continuous and active submarine surveillance of allied countries" and to the increase in the "safety of underwater infrastructure" such as cables in the waters around Iceland.[156] This was a reference to stepped up U.S. and NATO warnings about possible Russian sabotage of undersea cables and pipelines following its invasion of Ukraine. In addition, a plan for an infrastructure upgrade of a port facility in Helguvík near Keflavík to enable NATO warships to dock there was approved by the Icelandic government.[157]

The Ukraine War paralyzed the cooperation of the Arctic Eight in the Arctic Council and other Northern institutions. Only a week after the war started, the Western Arctic states announced what they termed a "pause in their participation in the Arctic Council" to protest Russia's "flagrant violation of the principles" of sovereignty and territorial integrity based on international law; they temporarily ceased participation in AC meetings or its subsidiary bodies.[158] Moreover, Russian Foreign Minister Sergey Lavrov was personally sanctioned by the United States, the EU, Canada, and the United Kingdom.[159] This did not, however, mean the end of the Arctic

Council—soon dubbed the Arctic Seven even if the Western Arctic states avoided that term[160]—which was described as having an "enduring value" for circumpolar cooperation. Moreover, in June 2022, they announced "a limited resumption" of their work in AC projects that had been approved by the Arctic Eight at the Reykjavík ministerial and that did not involve the participation of the Russian Federation.[161] Since the Arctic Council is only an intergovernmental forum and not based on an international treaty, it gave the Arctic Seven some leeway to cooperate as a group and to convene private, non-publicized meetings among themselves. Russia does not have a prominent role in many AC's projects, and it only leads one working group.[162] Nonetheless, any institutionalization of an Arctic Seven collaboration at the expense of Russia could jeopardize the Arctic Council, whose foundation was based on the equality of the Arctic Eight by increasing the likelihood of Russia's withdrawal from it.[163]

Russia, it turned out, allowed the transition from its chairmanship to Norway in May 2023 to go smoothly by agreeing to a save-facing formula: to convene an Arctic Council meeting in Moscow but allowing Senior Arctic officials of other AC states to attend it online. Since the late 2000s, the foreign ministers of Arctic states had usually attended such meetings, but since it was not required, their absence in Moscow could not be interpreted as a breach of any rules.[164] During its AC chairmanship, Russia stuck to its program of organizing over 40 Arctic events in ten Arctic regions, claiming that 28,000 people participated in them, but with foreign attendance limited to "friendly states" or those that have not followed a Western line on the Ukraine War, such as China, Belarus, India, Brazil, and Kazakhstan.[165]

When handing over the AC chairmanship to Norway, Lavrov did not suggest that Russia was planning to leave the Council. Yet, without mentioning Russia's own appetite for violent territorial conquests, he blamed the West for its military actions in Ukraine. He criticized Western states for suspending the "Council's full-scale activity under the absolutely far-fetched pretext that it related to the situation in Ukraine, which was provoked by the Western countries themselves."[166] The future of the AC depended on the continuation of a "civilized dialogue in the interests of preserving the Arctic as a territory of peace, stability and constructive cooperation" in line with the objectives put forward in the 2021 Ministerial Declaration in Reykjavík and the Strategic Plan of the Arctic Council until 2030.[167]

After Norway took over the reins of the AC, the gap between the Western Arctic countries and Russia widened, with no Arctic Council meetings taking place, not even among the Arctic Seven. What is more, the research activities of the working groups also became impaired. The Norwegians identified 55 AC research projects with non-Russian involvement, but there was minimum progress on their implementation.[168] In February 2024, the Russians stated that its continued AC membership "depended on the resumption of ... [its] activities and its compliance with Russian interests."[169] Russia had already reached out to other countries on the grounds that it should not lean on just one Arctic cooperation format with the involvement of various geographic parties. It is further underscored by their plan to develop an Arctic station on Svalbard with BRICS partners Brazil, India, China, and South Africa.[170] After Russia's threat, the Arctic Eight agreed on resuming the work of the Arctic working groups in a distant format. Whether this move signals the first step toward future Arctic Council cooperation remains to be seen.

The Ukraine War led to Russia's exclusion from participation in other Arctic-related subregional organizations that were created in the 1990s. In March 2022, the Nordic states and the EU suspended "activities involving Russia in the Barents Euro-Arctic cooperation, while reiterating their support for the institution and its work."[171] In addition, Norway and Iceland joined the EU in discontinuing all activities that involved Russia and Belarus (as an observer) in the Northern Dimension.[172] As for the Council of the Baltic Sea States, the members—which apart from the Nordic countries and the European Union also included Germany, Poland, and the Baltic states—went further by suspending Russia from the CBSS "until it is possible to resume cooperation based on respect for fundamental principles of international law."[173] In the Council's terms of reference, there is no provision for such enforced withdrawal, even if Russia's attack on a sovereign state can be interpreted as having violated its core principle of regional cooperation; after all, Ukraine has an observer status in the CBSS.[174]

The membership of the Baltic states and Poland, whose rhetoric against Russia, has often been more vocal than that of Western European states could also have played a role. In any case, apart from the Council of Europe—which had decided to expel Russia before the Russians withdrew on their own volition—the CBSS took a more hardline stance than other European regional organizations. The Russian government reacted badly to these actions, accusing the West of "anti-Russian hysteria" and claiming that without Russia the Barents Euro-Arctic Council and Northern Dimension would "lose their meaning."[175] Still, while it stopped short of severing its ties with the Northern Forum, it decided to withdraw from the CBSS and BEAC. When leaving the CBSS in May 2022, it argued that NATO and EU states had made "illegal and discriminatory decisions in violation of the rule of consensus" and turned the CBSS into "an instrument of anti-Russian politics."[176] In September 2023—when the Russians announced their departure from BEAC—they blamed the Western members for paralyzing the body's activities and for hindering the transfer of its rotating presidency to Russia.[177]

The Arctic Economic Council approached Russia's war against Ukraine differently or by holding a vote under the Russian chairmanship. Except for the Russian chair, the other executive committee members condemned the invasion. Yet the AEC continues to function, in a skeleton form, issuing press releases on issues relating to business and Arctic communities. After Norway took over the chairmanship from Russia, the new Norwegian AEC chair stated that there could be no business as usual, but the Council would continue its efforts to promote economic development in the Arctic.[178]

Despite the hardening of Iceland's foreign policy against Russia, it has been among those Arctic states that have not ruled out the future possibility of normalizing relations with Russia in the Arctic Council.[179] The United States, Canada, and Denmark initially expressed doubts that such cooperation could be restored,[180] but they agreed on the gradual resumption of the work of the working groups. There is, of course, an awareness among the Arctic Seven that Russia could try to delegitimize the Arctic Council by forming a rival Arctic regional organization with other more friendly states such as China.[181] Indeed, Russia concluded, in April 2023, an agreement on "maritime law enforcement" with China in the Arctic.[182] In other words, the absence of Russia from the Arctic Council could undermine pre-existing Arctic governance arrangements as well as the privileged position of the Arctic states.

Prolonged non-cooperation in the Arctic Council would not only harm intergovernmental relations in the region but also the interests of its inhabitants. Hence, it is not surprising that Indigenous groups have pressed for continued Arctic collaboration, pointing out that the social, economic, environmental, and health needs of their communities have little to do with the war in Ukraine or geopolitical rivalries.[183] Some were also critical of the Arctic Seven for excluding them from their unofficial meetings in 2022, prompting the former to meet with the latter in an attempt to mollify Indigenous opinion. The return to geopolitics in the Arctic could also affect the fight against climate change, with the buildup of greenhouse gases in the atmosphere being three to four times the global average. The loss of sea ice in the Arctic, the melting of the Greenland ice cap, and the thawing of permafrost are major drivers of climate change.[184]

It is too early to predict the medium- and long-term impact of Russia's military intervention in Ukraine on Arctic geopolitics and governance. After Finland's and Sweden's accession to NATO, a sharp dividing line has emerged between Russia and the seven other Western Arctic states. It is likely that NATO will increasingly view the Arctic as being part of its North Atlantic military posture and put more emphasis on the protection of the sea lines of communication. Cold War maritime strategies have already been revived for that purpose, and the stage could be set for a NATO-Russia military balancing across the region. Due to its location, Iceland would be part of any such geostrategic conjectures, which could also increase pressure for the reestablishment of a Cold War-style military presence on the island.

What is clear is that an era that has been driven by depoliticized cooperation narratives has ended. The outcome of the Ukraine War will determine whether—or to what degree—it will be possible to resume collaboration among the Arctic Eight. All the soft-law Arctic organizations that were created in the 1990s to deal with the Arctic and Russia's institutional integration with Western states—the AC, BEAC, ND, and CBSS—have ceased to function as originally intended. There is no hegemonic power, which is capable of constructing and policing a regional Arctic order. From a long-term perspective, it will be difficult to exclude Russia, which, as noted, makes up about half of the Arctic, from any Arctic governance projects. While Russia's foreign policy has targeted Ukraine, in particular, and the West, in general, it does not automatically mean that it will extend its armed confrontational stance to the Arctic.

Yet Russia's military presence in the Arctic plays a key role in defending the Kola Peninsula, with its second-strike nuclear assets. It is estimated that Russia has 58 nuclear submarines in its fleet, while the U.S. number is 64.[185] After Russia invaded Ukraine, it had to reduce its military posture in the Arctic because of pressures generated by the war. The U.S. Navy, on the other hand, is expanding as part of its own modernization drive and has begun building its largest and most advanced nuclear-powered ballistic missiles submarine. In NATO's 2022 strategic concept, there is only one reference to the "High North" as a challenge due to Russia's capability to disrupt Allied reinforcements and freedom of navigation across the North Atlantic. The use of the elastic Norwegian political term "High North" suggests that what is meant is the "GIUK gap."[186] NATO has also been warning against Russian mapping of—and capabilities to target—critical undersea infrastructure such as pipelines and cables that are crucial to communication systems.[187] It does not necessarily signify a full-scale return to a Cold War maritime posture—and may reflect awareness of gray zone and sub-threshold threats.

The Ukraine War represents, nonetheless, a watershed moment in world politics. Even if NATO's main focus is currently on the East of the Alliance, it has already had an impact on the North as well as other regions. This was underscored by the decision of the Icelandic Foreign Minister Gylfadóttir to suspend, in June 2023, Iceland's embassy operations in Moscow—without severing diplomatic relations—on the grounds that the level of commercial, cultural, and political relations between the two countries had reached an "all-time low."[188] How it came about is a bit murky. It was justified on utilitarian grounds—that a small country like Iceland could not maintain a full-fledged embassy without any practical functions. Ambassador Árni Þór Sigurðsson's time was up, having served in Moscow for three years, and he was slated for a move to Copenhagen. Still, a political motive also played a part because it was stated publicly that it depended on Russia's diplomatic conduct when normal relations could be restored.[189] This inconsistency was echoed in the press release about the decision.[190] Moreover, since the Icelandic Foreign Minister asked Russia to reciprocate by scaling back its diplomatic activities in Reykjavík, it meant that the controversial Russian Ambassador Mikhail Noskov—who had on occasions criticized the Icelandic Foreign Minister in undiplomatic terms because of her pro-Ukrainian stance—had to leave the country. The reaction of the Russian Foreign Ministry showed that the Russians viewed the closing of the embassy as an "unfriendly" act that destroyed "the entire range of Russian-Icelandic cooperation";[191] it would be taken into account "when building our ties with Iceland in the future," for "all anti-Russian actions would be followed by a corresponding reaction."[192]

Iceland was reminded by the Russians that "it is known that Russia had made a significant contribution to the formation of a sovereign Icelandic state, being one of the first to recognize its independence, and did much to protect Iceland's interests in difficult times for its people."[193] It is true that Russia was the third state—following the United States and the United Kingdom—to recognize the new status of Iceland as a republic following its abrogation of the Act of Union with Denmark in 1944.[194] Since Denmark recognized Iceland's full sovereignty in 1918—irrespective of the contractual relationship with the Danish King—there was no need to recognize it specifically as a republic; after all, a number of independent states had constitutional ties with another state. As a political symbolic act, however, this was important because successive Icelandic governments—constrained by the country's economic weakness—had refrained from seeking formal external recognition of Iceland as an independent state in the 1920s and 1930s. Moreover, after World War II, Russia sometimes played an important role in Iceland's economy through the barter trade relationship;[195] the prospective Russian loan during the financial crisis and the subsequent brief revival of the trade relationship, which lasted until the Crimea annexation, have also been discussed.

Whether the decision to close the Icelandic embassy in Moscow—which the Foreign Minister made unilaterally, and which was not met with approval by some government ministers[196]—will turn out to be consequential for Russian-Iceland relations remains to be seen. Gylfadóttir successor as Foreign Minister, the conservative Bjarni Benediktsson—the head of the Independence Party—seemed to be unsure of its wisdom, stating, in the fall of 2023, that he could not see that it had yielded any specific benefits.[197] Some Icelandic politicians and government officials disagreed with Gylfadóttir's step, which was not replicated by other countries, seeing a value in having a diplomatic representation in Moscow during the war and arguing that it could complicate the restoration of bilateral relations at a future date.[198] There is no means of knowing whether Russia will engage in memory politics and use delaying or punishing tactics when the question of normalizing diplomatic ties will be put back on the agenda. Iceland plays a minor role in Russian foreign and Arctic policies. Yet its renewed importance as a military base for the United States and NATO means that Russia cannot ignore its strategic position. The geopolitical uncertainties generated by the invasion of Ukraine mean that the future of the Icelandic-Russian relationship—as well as that of multilateral cooperation within the Arctic Council—will be contingent on a combination of factors: the outcome of the Ukraine War, internal Russian political developments, and the state of Western-Russian political relations, in general.

Notes

1. See Jason Dittmer, Sami Moisio, Alan Ingram, and Klaus Dodds. "Have you heard the one about the disappearing ice? Recasting Arctic geopolitics," *Political Geography* 30, no. 4 (2011): 202–214.
2. On the neo-realist-liberal institutionalism debate, see, for example, Charles Kegley, *Controversies in International Relations Theory: Realism and the Neoliberal Challenge* (New York: St. Martin's Press, 2015); Robert Keohane, *International Institutions and State Power: Essays in International Relations Theory* (New York: Routledge, 2020); Keohane and Joseph J. Nye, *Power and Interdependence*, 3rd. ed. (London: Longman, 2000); Kenneth N. Waltz, *Theory of International Relations* (Reading: Addison-Wesley, 1979); Robert O. Keohane, ed. *Neorealism and Its Critics* (New York: Columbia University Press, 1986); Filippo Andreatta and Mathias Koenig-Archibugi, "Which Synthesis? Strategies of Theoretical Integration and the Neorealist-Neoliberal Debate," *International Political Science Review* 31, no. 2 (2010): 207–227; John Mearsheimer, "Structural Realism," in *International Relations Theories: Discipline and Diversity*, ed. Tim Dunne, Milja Kurki, and Steve Smith (Oxford: Oxford University Press, 2007), 51–68; David A. Baldwin, ed. *Neo-realism and Neo-liberalism: The Contemporary Debate* (New York: Columbia University Press, 2003); Robert Jervis, "Realism, Neoliberalism, and Cooperation: Understanding the Debate," *International Security* 24, no. 1 (1999): 42–63.
3. See, for example, "The Ilulissat Declaration."
4. Interviews with Arctic Council officials, October 18, 2016.
5. See John Andreas Olsen, ed., *NATO and the North Atlantic: Revitalizing Collective Defence*, Whitehall Papers, Vol. 87 (London: RUSI, 2016); Expert Commission on Norwegian Security and Defense Policy, *Unified Effort* (Oslo: Norwegian Ministry of Defense, 2015), accessed November 15, 2023, https://www.regjeringen.no/globalassets/departementene/fd/dokumenter/unified-effort.pdf.
6. Carl Schmitt, *The Concept of the Political* (Chicago and London: The University of Chicago Press, 2007 [1932]).
7. See Jon Rahbek-Clemmensen, "The Ukrainian crisis moves north. Is Arctic conflict spill-over driven by material interests?" *Polar Record* 53, no. 1 (2017): 1–15.
8. James G. Foggo and Alarik Fritz, "NATO and the Challenge in the North Atlantic and the Arctic," in *Security in Northern Europe: Deterrence, Defence and Dialogue*, ed. Johan Andreas Olsen (London: RUSI, 2018), 121–128.

9. See Jørgen Staun, "Russia's strategy in the Arctic: cooperation, not confrontation," *Polar Record* 53, no. 3 (2017): 314.
10. See Icelandic Prime Minister's Office, report, *The Economic Impact of the Russian Counter-Sanctions on Trade between Iceland and the Russian Federation.*
11. Interview with a senior Icelandic foreign ministry official, November 3, 2015; see also an interview with Finance Minister Bjarni Benediktsson, the head of the Independence Party, "Bjarni hafði efasemdir frá upphafi" [Bjarni Benediktsson Harbored Doubts from the Start], *Morgunblaðið*, August 20, 2015, accessed June 10, 2023, https://www.mbl.is/frettir/innlent/2015/08/20/bjarni_hafdi_efasemdir_fra_upphafi/; see also Baldur Thorhallsson and Pétur Gunnarsson, "Iceland's alignment with the EU-US sanctions on Russia: autonomy versus dependence," *Global Affairs* 3, no. 3 (2017): 307–318.
12. "Bjarni hafði efasemdir frá upphafi"; "Sigmundur Davíð vill endurmeta EES-samninginn og ekki taka þátt í refsiaðgerðum 'blindandi'" [Sigmundur Davíð Gunnlaugsson Wants to Reevaluate the EEA-Agreement and Is against 'Blindingly' Taking Part in Economic Sanctions], *Kjarninn,* January 6, 2016, accessed September 13, 2022, https://kjarninn.is/frettir/2016-01-06-sigmundur-david-vill-endurmeta-ees-samninginn-og-ekki-taka-thatt-i-refsiadgerdum-blindandi/.
13. See "Sigmundur Davíð vill endurmeta EES-samninginn og ekki taka þátt í refsiaðgerðum 'blindandi.'"
14. Nikolai Wammen, Carl Hagland, Gunnar Bragi Sveinsson, Ine Eriksen Søreide, and Peter Hultqvist, "Við aukum norrænt samstarf á sviði varnarmála" [We Will Increase Nordic Defense Cooperation], April 10, 2015, accessed November 17, 2023, https://www.stjornarradid.is/efst-a-baugi/frettir/stok-frett/2015/04/10/Vid-aukum-norraent-samstarf-a-svidi-varnarmala/.
15. See *Skýrsla Gunnars Braga Sveinssonar utanríkisráðherra um utanríkis- og alþjóðamál*, March 2016; see also Foreign Minister Lilja Alfreðsdóttir's address—entitled "Brottför varnarliðsins—þróun varnarmála" [Icelandic National Security in the Post-IDF Era]—at a conference on the tenth anniversary of the departure of U.S. forces from Iceland, October 6, 2016, accessed September 30, 2023, https://www.stjornarradid.is/library/04-Raduneytin/Utanrikisraduneytid/PDF-skjol/Vardberg%2D%2D-raeda-utanri%CC%81kisra%CC%81dherra.pdf.
16. *Skýrsla Gunnars Braga Sveinssonar utanríkisráðherra um utanríkis- og alþjóðamál* [Report by Foreign Minister Gunnar Bragi Sveinsson on Foreign and International Affairs], *Þingtíðindi*, March 2014, accessed September 30, 2022, https://www.althingi.is/altext/143/s/0757.html; see also *Skýrsla Gunnars Braga Sveinssonar utanríkisráðherra um*

utanríkis- og alþjóðamál [Report by Foreign Minister Gunnar Bragi Sveinsson on Foreign and International Affairs], *Þingtíðindi*, March 2015.

17. It concerned the leak of millions of files from the world's fourth-largest offshore law firm, Mossack Fonseca. Sigmundur Davíð Gunnlaugsson, who was mentioned in the files due to his co-ownership with his wife of an offshore company, resigned as Prime Minister following mass public protests and pressure within his own Progressive Party.
18. Lilja Alfreðsdóttir, Brottför varnarliðsins—þróun varnarmála.
19. See Nataliya Marchenko, Odd Jarl Borch, Natalia Andreassen, Svetlana Kuznetsova, Valur Ingimundarson, and Uffe Jakobsen, "Navigation Safety and Risk Assessment Challenges in the High North," in *Marine Navigation and Safety of Sea Transportation,* 275; see also Ingimundarson and Halla Gunnarsdóttir, "The Icelandic Sea Areas and Activity Level up to 2025."
20. The member states are the following: Canada, Denmark, Finland, France, Germany, Iceland, the Netherlands, Norway, Russia, Sweden, the United Kingdom, and the United States.
21. See Arctic Coast Guard Forum, accessed February 22, 2024, https://www.arcticcoastguardforum.com/about-acgf.
22. See *Þjóðaröryggisstefna Íslands* [Icelandic National Security Policy], *Þingtíðindi*, April 14, 2016, accessed September 30, 2022, https://www.althingi.is/thingstorf/thingmalalistar-eftir-thingum/ferill/?ltg=145&mnr=327; see also Page Wilson and Auður H. Ingólfsdóttir, "Small State, Big Impact? Iceland's First National Security Policy."
23. See "Joint Declaration between the Department of Defense of the United States of America and the Ministry for Foreign Affairs of Iceland," June 29, 2016, accessed September 30, 2022, https://www.stjornarradid.is/library/04-Raduneytin/Utanrikisraduneytid/PDF-skjol/Joint-Declaration%2D%2DSigned-.PDF.
24. "Agreement between the United States of America and Iceland. Effected by the Exchange of Notes at Reykjavik, October 13 and 17, 2017," accessed September 30, 2023, https://www.state.gov/wp-content/uploads/2019/02/17-1017-Iceland-Defense-Coop-Notes.pdf.
25. Icelandic Parliament, "Svar utanríkisráðherra við fyrirspurn frá Andrési Inga Jónssyni um viðveru herliðs á Keflavíkurflugvelli [The Foreign Minister's [Guðlaugur Þór Þórðarson] Response to a Question by Andrés Ingi Jónsson, on [Foreign] Military Presence at Keflavík Airport]," November 14, 2018, accessed January 13, 2024, https://www.althingi.is/altext/149/s/0427.htm; see also "Voru 153 daga við kafbátaeftirlit á Íslandi [Conducted Submarine Surveillance for 153 Days]," *RUV*, November 14, 2018, accessed February 3, 2024, https://www.ruv.is/frettir/innlent/voru-153-daga-vid-kafbataeftirlit-a-islandi.

26. This policy tension is still present in Norway's 2020 Arctic policy. See Norwegian Ministry of Foreign Affairs, *Mennesker, muligheter og norske interesser i nord* [People, Opportunities and Norwegian Interests in the North], November 27, 2020, accessed September 30, 2022, https://www.regjeringen.no/no/dokumenter/meld.-st.-9-20202021/id2787429/; see also Andreas Østhagen, "What is the Point of Norway's new Arctic Policy?" The Arctic Institute, December 2, 2020, accessed September 30, 2022, https://www.thearcticinstitute.org/point-norway-new-arctic-policy/.
27. Guðlaugur Þór Þórðarson, "Arctic: Territory of Dialogue," address at the International Arctic Forum in Arkhangelsk, March 29, 2017, accessed September 30, 2022, https://www.stjornarradid.is/raduneyti/utanrikisraduneytid/utanrikisradherra/stok-raeda-utanrikisradherra/2017/03/29/Avarp-a-International-Arctic-Forum-i-Arkhangelsk/.
28. James Stavrividis, "The U.S. Aim at Russia with a Resurrected Navy Fleet," *Bloomberg*, May 16, 2018, accessed May 31, 2023, https://www.bloomberg.com/view/articles/2018-05-16/u-s-navy-takes-aim-at-russia-with-a-new-fleet.
29. "The U.S. Aim at Russia With a Resurrected Navy Fleet."
30. In his first annual report to the parliament on foreign and international affairs, Guðlaugur Þór Þórðarson did not single out the Arctic but placed much emphasis on NATO and Nordic security cooperation. See *Skýrsla Guðlaugs Þórs Þórðarsonar utanríkisráðherra um utanríkis- og alþjóðamál* [Report by Foreign Minister Guðlaugur Þór Þórðarsonar on Foreign and International Affairs], *Þingtíðindi*, May 2017, accessed September 30, 2022, https://www.althingi.is/altext/146/s/0671.html.
31. Guðlaugur Þór Þórðarson, "Back to the Future: The Geopolitical Centrality of the North Atlantic and the Arctic," address delivered at the Center for Strategic Studies, Washington, D.C., May 16, 2018, accessed November 15, 2022, https://www.stjornarradid.is/raduneyti/utanrikisraduneytid/utanrikisradherra/stok-raeda-utanrikisradherra/2018/05/16/Raeda-radherra-hja-hugveitunni-CSIS-i-Washington-DC/.
32. U.S. Department of Defense, *Summary of the 2018 National Defense Strategy: Sharpening the American Military Competitive Edge* (Washington D.C.: Department of Defense, 2018), accessed November 18, 2022, 3, https://dod.defense.gov/Portals/1/Documents/pubs/2018-National-Defense-Strategy-Summary.pdf.
33. See James G. Foggo and Alarik Fritz, "NATO and the Challenge in the North Atlantic and the Arctic," 121–128.
34. See, for example, Vahid Nick Pay and Harry Gray Calvo, "Arctic Diplomacy: A Theoretical Evaluation of Russian Foreign Policy in the High North," *Russian Politics*, 5, no. 1 (2020): 110–112.

35. State Council of the People's Republic of China, *China's Arctic Policy*, January 26, 2018, accessed September 30, 2021, https://english.www.gov.cn/archive/white_paper/2018/01/26/content_281476026660336.htm.
36. This approach is reminiscent of the Cold War debate during the Carter Administration (1976–1980), over whether to tie Soviet actions on issues, such as human rights, with the future of détente in Soviet-American relations. See David Skidmore, "Carter and the Failure of Foreign Policy Reform," *Political Science Quarterly* 108, no. 4 (1993/1994): 699–729; Gaddis Smith, *Morality, Reason, and Power: American Diplomacy in the Carter Years* (New York: Hill and Wang, 1986); Robert A. Strong, *Working in the World: Jimmy Carter and the Making of American Foreign Policy* (Baton Rouge: Louisiana State University Press, 2000).
37. Congressional Research Service, *Changes in the Arctic: Background and Issues for* Congress (updated February 1, 2021), 33, https://crsreports.congress.gov/product/pdf/R/R41153/177.
38. See Christopher R. Rossi, "Treaty of Tordesillas Syndrome: Sovereignty ad Absurdum and the South China Sea Arbitration," *Cornell International Law Journal*, 50, no. 2 (2017): 231–283; Oriana S. Mastro, "How China is bending the rules in the South China Sea"; Christine Elizabeth Macaraig and Adam James Fenton, "Analyzing the Causes and Effects of the South China Sea Dispute," *The Journal of Territorial and Maritime Studies*, 8, no. 2 (2021): 42–58.
39. See Marc Lanteigne, "The Rise (and Fall?) of the Polar Silk Road," *The Diplomat*, August 29, 2022, accessed June 2, 2023, https://thediplomat.com/2022/08/the-rise-and-fall-of-the-polar-silk-road/.
40. Oriana S. Mastro, "How China is bending the rules in the South China Sea," *The Interpreter*, February 17, 2021, accessed May 21, 2023, https://www.lowyinstitute.org/the-interpreter/how-china-bending-rules-south-china-sea.
41. See "Pompeo—Russia is 'aggressive' in the Arctic, China's work there also needs watching," *Reuters*, May 6, 2019, accessed September 30, 2022, https://www.reuters.com/article/us-finland-arctic-council/pompeo-russia-is-aggressive-in-arctic-chinas-work-there-also-needs-watching-idUSKCN1SC1AY.
42. See Mikkel Runge Olesen, "The end of Arctic exceptionalism? A review of the academic debates and what the Arctic prospects mean for the Kingdom of Denmark," *Danish Foreign Policy Review 2020* (Copenhagen: Danish Institute for International Studies), 103–127.
43. Interview with a Danish Arctic researcher, September 19, 2019.

44. Lon Strauss, "U.S. Marines and NATO's Northern Flank," *Arctic Review on Law and Politics* 13 (2022): 72–93, accessed November 23, 2024, https://arcticreview.no/index.php/arctic/article/view/3381/6326.
45. Julienne Smith and Jerry Hendrix, *Forgotten Waters: Minding the GIUK Gap.*
46. U.S. Department of State, "Remarks with Icelandic Minister of Foreign Affairs Ingibjörg Solrun Gisladottir," May 30, 2008, accessed February 3, 2024, https://2001-2009.state.gov/secretary/rm/2008/05/105447.htm.
47. See Valur Ingimundarson, *Uppgjör við umheiminn*, 50–54.
48. See comments by Prime Minister Katrín Jakobsdóttir, *Vísir*, September 4, 2019, accessed November 23, 2023, https://www.visir.is/g/20191909 09439/loftslagsvain-og-uppbygging-bandarikjahers-i-keflavik-storu-umraeduefnin-; see also the response from the non-committal response about Iceland's participation in the BRI from the Icelandic Ministry for Foreign Affairs to the webzine *Kjarninn*, September 5, 2019, accessed November 18, 2022, https://kjarninn.is/skyring/2019-09-05-engin-endanlega-afstada-verid-tekin-til-thatttoku-i-belti-og-braut/.
49. U.S. Air Forces in Europe and Air Forces Africa Public Affairs, "U.S. Air Force B-2 Spirit stealth bomber aircraft arrive in Iceland for ally, partner training," August 24, 2021, accessed February 6, 2024, https://www.usafe.af.mil/News/Press-Releases/Article/2743096/us-air-force-b-2-spirit-stealth-bomber-aircraft-arrive-in-iceland-for-ally-part/.
50. See U.S. Embassy in Copenhagen, "Statement of Intent on Defense Investments in Greenland," September 17, 2018, accessed December 13, 2022, https://twitter.com/usembdenmark/status/10416952406866 32960?lang=en.
51. See Valur Ingimundarson, "Framing the national interest: the political uses of the Arctic in Icelandic foreign and domestic policies."
52. See "Obama and Nordic Nations Discuss Russia," *New York Times*, May 13, 2016, accessed May 25, 2023, https://www.nytimes.com/video/us/politics/100000004405045/obama-and-nordic-nations-discuss-russia.html, May 13, 2016; "Obama Warms Up to Nordic Leaders," *New York Times*, May 13, 2016, accessed May 25, 2023, https://www.nytimes.com/2016/05/14/world/europe/obama-warms-to-nordic-heads-of-state.html.
53. About the fading interest of the Nordic countries in cultivating relations with China, see, for example, Andreas Bøje Forsby, "Falling out of Favor: How China Lost the Nordic Countries," *The Diplomat*, June 24, 2022, accessed May 25, 2023, https://thediplomat.com/2022/06/falling-out-of-favor-how-china-lost-the-nordic-countries/.
54. Interview with an Icelandic academic, February 3, 2024.

55. Ingi Freyr Vilhjálmsson, "Kínverska ríkið setur 700 til 800 milljónir í rannsóknarmiðstöð um norðurljósin" [China Invests 700–800 Million [Icelandic Kronor] in an Aurora Observatory], *Heimildin*, March 6, 2023, accessed April 10, 2023. https://heimildin.is/grein/16852/.
56. Interview with an Icelandic science official, December 16, 2022.
57. Melody Schreiber, "A new China-Iceland Arctic science observatory is already expanding its focus," *Arctic Today*, October 31, 2018, accessed July1,2023,https://www.arctictoday.com/new-china-iceland-arctic-science-observatory-already-expanding-focus/.
58. See Þorsteinn Gunnarsson and Egill Þór Níelsson, "An Icelandic Perspective," in *Nordic-China Cooperation*, 87–93.
59. "Kínverska ríkið setur 700 til 800 milljónir í rannsóknarmiðstöð um norðurljósin"; interview with a Icelandic science official December 16, 2022.
60. Email communications from Gunnlaugur Björnsson, a Research Scientist at the University of Iceland, to Valur Ingimundarson, February 8 and 10, 2023.
61. Interview with an Icelandic official, October 20, 2024.
62. See Ingi Freyr Vilhjálmsson, "Þingmaður spyr Katrínu um eftirlit með kínversku rannsóknarmiðstöðinni" [A Parliamentarian Asks Katrín [Jakobsdóttir] about how the Chinese Research Center Is Monitored], *Heimildin*, April 3, 2023, accessed April 10, 2023, https://heimildin.is/grein/17311/; Ingi Freyr Vilhjálmsson, "Önnur ríki hafa áhyggjur af norðurljósamiðstöð Kína á Íslandi" [Other Countries Are Worried about the Aurora Observatory], *Heimildin*, April 1, 2023, accessed April 10, 2023, https://heimildin.is/grein/17281/; Vilhjálmsson, "NATO hefur lýst áhyggjum af rannsóknamiðstöð Kína um norðurljósin" [NATO Has Expressed Worries about China's Northern Lights Research Center], *Heimildin*, March 28, 2023, accessed April 10, 2023, https://heimildin.is/grein/17192/.
63. See, for example, "China Begins to Revive Arctic Scientific Ground Projects After Setbacks," *Voice of America*, December 5, 2022, accessed January 4, 2023, https://www.voanews.com/a/china-begins-to-revive-arctic-scientific-ground-projects-after-setbacks/.
64. See Icelandic Ministry for Foreign Affairs. *Norðurljós. Skýrsla starfshóps um efnahagstækifæri á norðurslóðum* [Northern Lights, a Working Group Report about Economic Opportunities in the Arctic], May 2021, accessed July 5, 2023, https://www.stjornarradid.is/library/04Raduneytin/Utanrikisraduneytid/PDF-skjol/Nor%C3%B0urlj%C3%B3s%20-%20efnahagst%C3%A6kif%C3%A6ri%20%C3%A1%20nor%C3%B0ursl%C3%B3%C3%B0um_WEB.pdf.
65. Interview with a high-ranking Icelandic Arctic official, March 12, 2021.

66. Letter, John Moloolenar and Raja Krishnamoorthi to Antony Blinken and Lloyd Austin, October 16, 2024, accessed October 25, 2024, https://selectcommitteeontheccp.house.gov/sites/evo-subsites/selectcommitteeontheccp.house.gov/files/evo-media-document/10.16.24_PRC%20dual%20use%20research%20in%20the%20Arctic; see also "Áhyggjur af norðurljósarannsóknum " [Worries about aurora reseach], Morgunblaðið, October 19, 2024, accessed October 25, 2024, https://www.mbl.is/greinasafn/grein/1873365/?item_num=2&dags=2024-10-19&t=515572980&_t=1729857459.4462316.
67. The biweekly newspaper *Heimildin*, accessed June 1, 2023, https://heimildin.is/.
68. The name of the parliamentarian is Andrés Ingi Jónsson.
69. Björn Bjarnason, "The China Stigma Must Be Erased," November 28, 2022, accessed June 12, 2023, https://www.bjorn.is/dagbok/afma-verdur-kinastimpilinn.
70. See "Íslensk stjórnvöld hafa ekkert eftirlit eða aðkomu að rannsóknarmiðstöð Kína [The Icelandic Government Does Not Monitor and Has No Involvement in China's Research Center]," *Heimildin*, June 9, 2023, accessed June 11, 2023, https://heimildin.is/grein/18033/islensk-stjornvold-hafa-ekkert-eftirlit-eda-adkomu-ad-rannsoknarmistod-kina/.
71. Interview with a high-ranking Arctic official, February 6, 2024.
72. About Chinese investments in Iceland, see Egill Þór Níelsson and Guðbjörg Ríkey Th. Hauksdóttir, "Kina, investeringer og sikkerhetspolitikk: Politikk og perspektiver i Norden—Island" [China, Investments and Security Policy: Politics and Perspectives in the Nordics—Iceland], *Internasjonal Politikk* 78, no. 1 (2020): 68–78; see also OECD, "OECD Benchmark Definition of Foreign Direct Investment—Fourth Edition," 2008, 17, accessed November 18, 2022, https://www.oecd.org/daf/inv/investmentstatisticsandanalysis/40193734.pdf.
73. See, for example, Guðlaugur Þór Þórðarson, "Iceland-China relations will continue to strengthen," *China Daily*, September 6, 2018, accessed December 30, 2022, https://www.chinadaily.com.cn/a/201809/06/WS5b90702ba31033b4f465477b.html.
74. *Þingsályktun um stefnu Íslands í málefnum norðurslóða.*
75. A German engineering company, Bremenports, has invested in preliminary research on the possibility of building such a deep-water port. See "Bremenports to Develop New Deep-Water Port in Iceland," *World Maritime News*, May 27, 2015, accessed November 30, 2022, https://worldmaritimenews.com/archives/275015/bremenports-to-develop-new-deep-water-port-in-iceland/; see also "Stærsta verkefni Íslandssögunnar. Hvað er að gerast í Finnafirði?" [The Greatest Project in

the History of Iceland. What Is Happening in Finnafjörður?], *Kjarninn,* March 21, 2021, accessed July 10, 2023, https://kjarninn.is/skyring/staersta-verkefni-islandssogunnar-hvad-er-ad-gerast-i-finnafirdi/.

76. "Drög að frumvarpi til laga um breytingu á varnarmálalögum nr. 34/2008 (á öryggissvæði o.fl.)" [A Draft of a Legislative Change in the Defense Law], February 26, 2021, accessed December 30, 2022, https://samradsgatt.island.is/oll-mal/$Cases/Details/?id=2932.
77. "Katrín segir nýja herstöð ekki koma til greina" [PM Katrín Jakobsdóttir Says That a New Military Base Is out of the Question], *Vísir*, November 3, 2022, accessed January 22, 2022, https://185.21.17.249/g/2020 2032771d/katrin-segir-nyja-herstod-ekki-koma-til-greina.
78. Interviews with a high-level Icelandic government official and defense official, February, 5, 2019.
79. Guðlaugur Þór Þórðarson, "New Geo-Political Reality in the West Nordic Area," a conference speech at the Arctic Circle Assembly, Reykjavík, October 20, 2018, accessed December 30, 2022.
80. "New Geo-Political Reality in the West Nordic Area."
81. *Skýrsla utanríkisráðherra um utanríkis- og alþjóðamál* [Report by the Minister for Foreign Affairs on Foreign and International Affairs], April 2019, 10, 17, accessed November 30, 2022, https://www.stjornarradid.is/efst-a-baugi/frettir/stok-frett/2019/04/29/Skyrsla-utanrikisradherra-um-utanrikis-og-althjodamal-logd-fyrir-Althingi/.
82. *Skýrsla utanríkis- og þróunarsamvinnuráðherra um utanríkis- og alþjóðamál* [Report by the Minister for Foreign Affairs and Development Cooperation] on Foreign and International Affairs, May 2020, 9, 79, accessed December 30, 2022, https://www.stjornarradid.is/gogn/rit-og-skyrslur/stakt-rit/2020/05/07/Skyrsla-utanrikis-og-throunarsamvinnuradherra-um-utanrikis-og-althjodamal-2020/; *Skýrsla utanríkis- og þróunarsamvinnuráðherra um utanríkis- og alþjóðamál* [Report by the Minister for Foreign Affairs and Development Cooperation on Foreign and International affairs], May 2021, 25, 81, accessed September 30, 2022, https://www.stjornarradid.is/efst-a-baugi/frettir/stok-frett/2021/05/05/Skyrsla-utanrikis-og-throunarsam vinnuradherra-um-utanrikis-og-althjodamal-2021/.
83. "Skref í átt að enn betri samvinnu: Fyrsti sendimaður Grænlands á Íslandi tekinn til starfa" [A Step Towards Deeper Cooperation: The First Greenlandic Representative Has Assumed His Functions], *Morgunblaðið,* January 3, 2019, accessed November July 20, 2024, https://www.mbl.is/greinasafn/grein/1709573/.
84. Icelandic Ministry for Foreign Affairs, *Samstarf Grænlands og Íslands á nýjum Norðurslóðum. Tillögur Grænlandsnefndar utanríkis- og þróunarsamvinnuráðherra* [Cooperation between Greenland and Iceland in the

New Arctic. Proposals by the Greenland Committee of the Minister for Foreign Affairs and International Development Cooperation], December 2020, accessed February 11, 2023, https://www.stjornarradid.is/library/02-Rit%2D%2Dskyrslur-og-skrar/Samstarf%20Gr%c3%a6nlands%20og%20%c3%8dslands%20%c3%a1%20n%c3%bdjum%20Nor%c3%b0ursl%c3%b3%c3%b0um%20-%20Copy%20(1).pdf.

85. *Samstarf Grænlands og Íslands á nýjum Norðurslóðum*, 37.
86. As Page Wilson has pointed out, the legal and political complexities of the Greenlandic independence question have often been downplayed by politicians and scholars. But if the Greenlanders opt for independence—which would mean the end of the annual block grant from Denmark in the amount of more than half a billion U.S. dollars or about 20% of Greenland's GDP and more than half of its budget—it is hard to see the Danes resisting such a decision given the contractual commitments they have already made; see Wilson, "An Arctic 'cold rush'? Understanding Greenland's (in)dependence question," *Polar Record* 53, no. 5 (2017): 512–519.
87. *Samstarf Grænlands og Íslands á nýjum Norðurslóðum*, 27–28, 30–31, 33.
88. *Samstarf Grænlands og Íslands á nýjum Norðurslóðum*, 19.
89. *Samstarf Grænlands og Íslands á nýjum Norðurslóðum*, 33.
90. *Samstarf Grænlands og Íslands á nýjum Norðurslóðum*, 14.
91. *Samstarf Grænlands og Íslands á nýjum Norðurslóðum*, 33.
92. *Þingsályktunartillaga um aukið samstarf Íslands og Grænlands* [A Parliamentary Resolution on Increased Cooperation between Iceland and Greenland], May 31, 2021, accessed February 6, 2024, https://www.althingi.is/altext/151/s/1560.html.
93. *Þingsályktunartillaga um aukið samstarf Íslands og Grænlands.*
94. High-ranking Icelandic official, interview by author, April 12, 2023.
95. "Bæta í samstarf Grænlands og Íslands" [The Expansion of Greenlandic-Icelandic Relations], *Morgunblaðið*, September 23, 2023, accessed July 5, 2023, https://www.mbl.is/frettir/innlent/2021/09/23/baeta_i_samstarf_graenlands_og_islands/.
96. Maria Ackrén and Rasmus Leander Nielsen, "The First Foreign- and Security Policy Opinion Poll in Greenland," Ilisimatusarfik/University of Greenland and Konrad Adenauer Stiftung, February 2021, accessed February 11, 2023, https://uni.gl/media/6762444/fp-survey-2021-ilisimatusarfik.pdf.
97. See "Merkilegt skref í samstarfi Íslands og Grænlands" [A Noteworthy Step in Icelandic-Greenlandic Cooperation], *Fréttablaðið*, October 13, 2022, accessed February 11, 2023, https://www.frettabladid.is/frettir/merkilegt-skref-i-samstarfi-islands-og-graenlands/.

98. Icelandic Prime Minister's Office, "Declaration of Cooperation between the Prime Minister of Greenland and the Prime Minister of Iceland on Future Cooperation between Greenland and Iceland," October 13, 2022, accessed February 11, 2023, https://www.government.is/library/01-Ministries/Prime-Ministrers-Office/Declaration%20of%20cooperation.pdf.
99. "Declaration of Cooperation between the Prime Minister of Greenland and the Prime Minister of Iceland on Future Cooperation between Greenland and Iceland."
100. Address by U.S. Secretary of State Mike Pompeo at the Arctic Council Ministerial Meeting in Rovaniemi.
101. Address by U.S. Secretary of State Mike Pompeo at the Arctic Council Ministerial Meeting in Rovaniemi.
102. Address by U.S. Secretary of State Mike Pompeo at the Arctic Council Ministerial Meeting in Rovaniemi.
103. Address by U.S. Secretary of State Mike Pompeo at the Arctic Council Ministerial Meeting in Rovaniemi.
104. Guðlaugur Þór Þórðarson, "Statement by the Minister of Foreign Affairs of Iceland," Arctic Council Ministerial Meeting, Rovaniemi, May 7, 2019, accessed January 25, 2024, https://oaarchive.arctic-council.org/handle/11374/2402?show=full.
105. Guðlaugur Þór Þórðarson, "Presentation of the [Arctic Council] Icelandic Chairmanship Program," Arctic Council Ministerial Meeting, Rovaniemi, May 7, 2019, accessed January 25, 2024, https://oaarchive.arctic-council.org/handle/11374/2397; see also Þórðarson, "Opening Address at the seminar 'Doing Business in the Arctic,'" American-Icelandic Chamber of Commerce, Washington, D.C., May 23, 2019, accessed January 25, 2022, https://www.stjornarradid.is/raduneyti/utanrikisraduneytid/utanrikisradherra/fyrri-radherrar/stok-raeda-fyrrum-radherra/2019/05/23/Opnunaravarp-a-malthingi-Islensk-Ameriska-vidskiptaradsins-Doing-Business-in-the-Arctic/.
106. Arctic Economic Council, "Banks banning investments in the Arctic will hurt even more now," May 5, 2022, accessed January 23, 2024, https://arcticeconomiccouncil.com/news/banks-banning-investments-in-the-arctic-will-hurt-even-more-now/.
107. See "Why the Arctic Economic Council is Needed: The People of the Arctic Deserve Our Full Attention and Opportunity to Develop," *High North News*, May 31, 2023, accessed May 31, 2023, https://www.highnorthnews.com/en/people-arctic-deserve-our-full-attention-and-opportunity-develop.
108. "First Joint Meeting between the Arctic Council and the Arctic Economic Council," Arctic Council, October 19, 2019, accessed May 31, 2023,

https://arctic-council.org/news/first-joint-meeting-between-the-arctic-economic-council/.

109. See Margrét Cela and Pia Hansson, "A challenging chairmanship in turbulent times."
110. Interview with a high-ranking Icelandic Arctic official, March 5, 2021.
111. Arctic Council, *Arctic Council Strategic Plan 2021–2030*, May 20, 2021, accessed January 25, 2024, https://oaarchive.arctic-council.org/handle/11374/2601.
112. "Reykjavík Declaration," Arctic Council, May 20, 2021, accessed January 25, 2024, https://oaarchive.arctic-council.org/handle/11374/2600.
113. "Outlines of the Statement by Sergey Lavrov at the 12th Ministerial Meeting of the Arctic Council," Reykjavík, May 20, 2021, accessed January 22, 2024, https://oaarchive.arctic-council.org/bitstream/handle/11374/2683/Statement%20by%20Minister%20S.Lavrov.pdf?sequence=1&isAllowed=; https://apnews.com/article/donald-trump-iceland-europe-russia-middle-east-24e823ef46207bfbe4c841e1b4a864f0y.
114. "Outlines of the Statement by Sergey Lavrov"; see also Kathryn C. Lavelle, "Regime, Climate, and Region in Transition: Russian Participation in the Arctic Council," *Problems of Post-Communism* 69, no. 4–5 (2022): 345–357.
115. See Daniel McVicater, "How the Russia-Ukraine War Challenges Arctic Governance," Council of Foreign Relations, May 10, 2022, accessed January 22, 2024, https://www.cfr.org/blog/how-russia-ukraine-war-challenges-arctic-governance.
116. "Top US, Russia diplomats spar firmly but politely in Iceland," *AP*, May 20, 2021, accessed January 22, 2024, https://apnews.com/article/donald-trump-iceland-europe-russia-middle-east-24e823ef46207bfbe4c841e1b4a864f0.
117. "Vill liðka fyrir bættum samskiptum" [Wants to Facilitate Improved Relations], *Fréttablaðið*, April 16, 2021, accessed January 22, 2024, https://frettabladid.overcastcdn.com/documents/210416.pdf.
118. "Ísland fest sig í sessi sem miðstöð norðurslóða" [The Consolidation of Iceland as a Central Arctic Hub], *Fréttablaðið*, May 21, 2021, accessed January 22, 2024, https://www.frettabladid.is/frettir/olafur-island-fest-sig-i-sessi-sem-midstod-nordursloda/.
119. "Ísland fest sig í sessi sem miðstöð norðurslóða."
120. "Ísland fest sig í sessi sem miðstöð norðurslóða."
121. See, for example, David Edmonds and John Eidinow, *Bobby Fischer Goes to War* (New York: Harper Publishers, 2004); Harold C. Schonberg, "Cold War in the World of Chess," *New York Times*, September 27, 1981, accessed February 6, 2024, https://www.nytimes.com/1981/09/27/

magazine/cold-war-in-the-world-of-chess.html; "Kissinger Phone Call to Fischer Disclosed," *New York Times,* July 18, 1972, accessed February 10, 2024, https://www.nytimes.com/1972/07/18/archives/kissinger-phone-call-to-fischer-disclosed.html.

122. *Þingsályktun um stefnu Íslands í málefnum norðurslóða.*
123. *Þingsályktun um stefnu Íslands í málefnum norðurslóða.*
124. Arctic Council, "New Observer: West Nordic Council," September 11, 2017, https://arctic-council.org/news/new-observer-west-nordic-council/.
125. *Þingsályktun um stefnu Íslands í málefnum norðurslóða.*
126. *Þingsályktun um stefnu Íslands í málefnum norðurslóða.*
127. Interview with an Icelandic official, February 23, 2024. See also Joji Morshita, "The Arctic Five-plus-Five process on central Arctic Ocean fisheries negotiations: Reflecting the interests of Arctic and non-Arctic actors," in *Emerging Legal Orders in the Arctic: The Role of Non-Arctic Actors*, ed. Akiho Shibata, Leilei Zou, Nikolas Sellheim, and Marzia Scopelliti, 109–131.
128. See Jóhann Sigurjónsson, "Icelandic Perspectives on the Agreement to Prevent Unregulated High Seas Fisheries in the Central Arctic Ocean," *The Yearbook of Polar Law Online* 12, no. 1 (2020): 268–284; see also Arnór Snæbjörnsson, "Nýr samningur um að koma í veg fyrir stjórnlausar úthafsveiðar í miðhluta Norður-Íshafsins" [New Agreement on the Prevention of Irregular Fishing on the High Seas in the Central Arctic Ocean], *Úlfljótur*, January 23, 2019, accessed February 20, 2024, https://ulfljotur.com/2019/01/23/nyr-samningur-um-ad-koma-i-veg-fyrir-stjornlausar-uthafsveidar-i-midhluta-nordur-ishafsins/; Sigurveig Þórhallsdóttir, "Norður-Íshafssamningurinn" [The Central Arctic Ocean Agreement] (MA thesis, University of Iceland, 2020).
129. Interview with a Canadian official, December 3, 2014.
130. "Iceland protests five-nation fishing deal in Arctic," *News24*, July 24, 2015, accessed January 15. 2024, https://www.news24.com/news24/green/news/iceland-protests-five-nation-fishing-deal-in-arctic-20150724.
131. Icelandic Ministry for Foreign Affairs, "Vegna '5-ríkja samráðs' um fiskveiðar í Norður-Íshafi" [On "Five-State Consultation" on Fishing in the Arctic Ocean], July 23, 2015, accessed February 18, 2024, https://www.stjornarradid.is/efst-a-baugi/frettir/stok-frett/2015/07/23/Vegna-5-rikja-samrads-um-fiskveidar-i-Nordur-Ishafi-nbsp/; see also "Iceland protests five-nation fishing deal in Arctic."
132. The Foreign Minister stressed that since Iceland was one of the largest fishing states in the world—and more dependent on fishing than any of the other Arctic states—it had a say over future regulation of fishing in international waters. Claiming that Iceland had always supported science-

based regional cooperation, he made the point that it was a member of the North East Atlantic Fisheries Commission (NEAFC), Northwest Atlantic Fisheries Commission (NAFO), and the International Council for the Exploration of the Seas (ICES). Finally, he argued that the 1995 UN Fish Stocks Agreement on the conservation and management of straddling fish stocks and migratory stocks covered the Arctic Ocean as well as other parts of the world. See "Vegna '5-ríkja samráðs' um fiskveiðar í Norður-Íshafi."

133. See Jon Rahbek-Clemmensen and Gry Thomasen, "How has Arctic coastal state cooperation affected the Arctic Council?" 6; see also Erik J. Molenaar, "International Regulation of Central Arctic Ocean Fisheries," in *Challenges of the Changing Arctic: Continental Shelf, Navigation, and Fisheries*, ed. Myron H. Nordquist, John Norton Moore, and Ronán Long (Leiden: Brill Nijhoff, 2016), 429–463; Erik J. Molenaar, "The CAOF Agreement: Key Issues of International Fisheries Law." In *New Knowledge and Changing Circumstances in the Law of the Sea*, ed. Tomas Heidar, (Leiden/Boston: Brill Nijhoff, 2020), 446–476.
134. See Jóhann Sigurjónsson, "Icelandic Perspectives on the Agreement to Prevent Unregulated High Seas Fisheries in the Central Arctic Ocean," 270.
135. "How has Arctic coastal state cooperation affected the Arctic Council?" 6.
136. "How has Arctic coastal state cooperation affected the Arctic Council?" 5.
137. Yet, when it comes to the management of migratory and transboundary fish stock in the Arctic region, Iceland had been keen on fighting any attempts to establish an international fishery management organization, which could limit its own influence as a major fishing state. Iceland has stated its opposition to any regional fishing management organizations (RFMOs), which excludes states with vested regional interests, with a say over the management of migratory, transboundary, and straddling fish stocks. On the other hand, it has been not opposed to regulatory measures designed to promote sustainability—in general terms—in the Arctic. Interview with a high-ranking Arctic official, January 20, 2017.
138. See Icelandic Ministry for Foreign Affairs, "Auglýsing um samning um að koma í veg fyrir stjórnlausar úthafsveiðar í miðhluta Norður-Íshafsins" [Announcement on the Agreement on the Prevention of Unregulated High Seas Fishing in the Central Arctic Ocean], June 22, 2021, https://www.stjornarradid.is/library/03-Verkefni/Utanrikismal/Lagamal/C-deildar-atak/16_2021_Augl%C3%BDsing%20um%20samning%20um%20a%C3%B0%20koma%20%C3%AD%20veg%20fyrir%20stj%C3%B3rnlausar%20%C3%BAthafsvei%C3%B0ar%20%C3%AD%20mi%C3%B0hluta%20Nor%C3%B0ur-%C3%8Dshafsins..pdf
139. *Tillögur nefndar um endurskoðun á stefnu Íslands í málefnum norðurslóða.*

140. *Tillögur nefndar um endurskoðun á stefnu Íslands í málefnum norðurslóða.*
141. See, for example, Rauna Kuokkanen, "'To See What State We Are In': First Years of the *Greenland Self-Government Act* and the Pursuit of Inuit Sovereignty," *Ethnopolitics* 16, no. 2 (2017): 184; see also Dorothée Céline Cambo, "Disentangling the conundrum of self-determination and its implications in Greenland," *Polar Record* 56 (e3): 1–10. https://doi.org/10.1017/S0032247420000169.
142. *Tillögur nefndar um endurskoðun á stefnu Íslands í málefnum norðurslóða.*
143. *Tillögur nefndar um endurskoðun á stefnu Íslands í málefnum norðurslóða.*
144. "Agreement on the Platform for the Coalition Government of the Independence Party, the Left Green Movement and the Progressive Party," November 28, 2021, accessed February 11, 2024, https://www.stjornarradid.is/rikisstjorn/stjornarsattmali/.
145. Icelandic Prime Minister's Office and the Ministry for Foreign Affairs, "Ísland fordæmir innrás Rússa í Úkraínu" [Iceland Condemns the Invasion of Russia into Ukraine], February 24, 2022, accessed February 11, 2024, https://www.stjornarradid.is/efst-a-baugi/frettir/stok-frett/2022/02/24/Island-fordaemir-innras-Russa-i-Ukrainu/; "Innrás Rússlands í Úkraínu—viðbrögð íslenskra stjórnvalda: íslensk stjórnvöld fordæma ólögmæta innrás Rússlands og lýsa yfir algjörum stuðningi við Úkraínu" [Russia's Invasion of Ukraine—Iceland's Reaction: The Icelandic Government Condemns Russia's Illegal Invasion and Declares Full Support for Ukraine], the Icelandic Government, October 6, 2022, accessed February 9, 2024, https://www.stjornarradid.is/verkefni/utanrikismal/strid-i-ukrainu-vidbrogd-islenskra-stjornvalda/.
146. "Ræða utanríkisráðherra [Þórdís Kolbrún Reykfjörð Gylfadóttir] á Helsinki Security Forum á þingi Hringborðs norðurslóða."
147. "Ræða utanríkisráðherra [Þórdís Kolbrún Reykfjörð Gylfadóttir] á Helsinki Security Forum á þingi Hringborðs norðurslóða."
148. "Ræða utanríkisráðherra [Þórdís Kolbrún Reykfjörð Gylfadóttir] á Helsinki Security Forum á þingi Hringborðs norðurslóða."
149. "Ræða utanríkisráðherra [Þórdís Kolbrún Reykfjörð Gylfadóttir] á Helsinki Security Forum á þingi Hringborðs norðurslóða."
150. "Mikil umsvif og náin samvinna" [Extensive [military] Activities and Close Cooperation], *Morgunblaðið*, April 13, 2023, accessed April 19, 2023, https://www.mbl.is/mogginn/bladid/grein/1833634/?t=380088780&_t=1681913551.4705353.
151. See, for example, "Föst viðvera herliðs á Íslandi hafi fælingarmátt" [A Permanent Military Presence in Iceland Serves a Deterrence Function], *Morgunblaðið*, March 6, 2023, accessed April 19, 2022, https://www.mbl.is/frettir/innlent/2022/03/06/fost_vidvera_herlids_a_islandi_hafi_faelingarmatt/; "Skoða ætti endurkomu hersins" [The Return of the

[U.S.] Military Should Be Explored], *Morgunblaðið*, April 7, 2023, accessed April 19, 2023, https://www.mbl.is/mogginn/bladid/netgreinar/2023/04TT/07/skoda_aetti_endurkomu_hersins/.

152. See Arnór Sigurjónsson, *Íslenskur her. Breyttur heimur, nýr veruleiki* [An Icelandic Military: New Challenges in a Changed World] (Reykjavík: Arnór Sigurjónsson, 2023).
153. And the Defense Agreement with the United States as the cornerstone of its defense policy.
154. See "Ekki mistök að loka varnarstöðinni" [Not a Mistake to Close the Defense Base], *Morgunblaðið*, April 20, 2023, accessed April 20, 2023, https://www.mbl.is/frettir/innlent/2023/04/14/ekki_mistok_ad_loka_varnarstodinni/; see also "Íslenskur her ekki tímabær" [An Icelandic Military Not Timely], *Morgunblaðið*, March 7, 2023, accessed April 20, 2023, https://www.mbl.is/frettir/innlent/2023/03/07/islenskur_her_ekki_timabaer/.
155. Interview with a U.S. Department of Defense official, April 10, 2018.
156. See Icelandic Ministry for Foreign Affairs, "Service visits of US submarines authorized," April 18, 2023, accessed June 3, 2023, https://www.government.is/diplomatic-missions/embassy-article/2023/04/18/Service-visits-of-US-submarines-authorised/#:~:text=The%20Minister%20for%20Foreign%20Affairs,supplies%20and%20exchange%20crew%20members.
157. See "Fyrirhugað að hafa herskip í Helguvík" [The Docking of Warships Planned in Helguvík], *Morgunblaðið*, June 9, 2023, accessed June 11, 2023, https://www.mbl.is/mogginn/bladid/grein/1837704/.
158. U.S. Department of State, "Joint Statement on Arctic Council Following Russia's Invasion of Ukraine" [The United States, Canada, Norway, Denmark, Finland, Sweden, and Iceland], March 3, 2022, accessed January 24, 2023, https://www.state.gov/joint-statement-on-arctic-council-cooperation-following-russias-invasion-of-ukraine/.
159. See Daniel McVicater, "How the Russia-Ukraine War Challenges Arctic Governance," Council of Foreign Relations, May 10, 2022, accessed January 22, 2023, https://www.cfr.org/blog/how-russia-ukraine-war-challenges-arctic-governance.
160. Interview with high-ranking Icelandic official, interview by author, April 12, 2023.
161. U.S. Department of State, "Joint Statement on Limited Resumption of Arctic Council Cooperation [The United States, Canada, Norway, Denmark, Finland, Sweden, and Iceland]," June 8, 2022, accessed January 24, 2023, https://www.state.gov/joint-statement-on-limited-resumption-of-arctic-council-cooperation/.

162. See Timo Koivurova, "The Arctic Council can continue without Russia," *Arctic Today,* March 10, 2022, accessed January 24, 2023, https://www.arctictoday.com/the-arctic-council-can-continue-without-russia/.
163. See "Russia's future in Arctic Council depends on whether its activity meets Russian interests—Russian Foreign Ministry," *Interfax,* February 5, 2024, accessed February 7, 2024, https://interfax.com/newsroom/top-stories/99094/; see also Timo Koivurova and Ahio Shibata, "After Russia's invasion of Ukraine in 2022: Can we still cooperate with Russia in the Arctic?" *Polar Record* 59, e12 (2023): 1–9, https://doi.org/10.1017/S0032247423000049.
164. Interview with a high-ranking Icelandic official, April 12, 2023; interview with a high-ranking Arctic official, interview by author, February 6, 2024.
165. "Lavrov: 'The Arctic Council's Future Depends on Whether a Civilized Dialogue Can Continue," *High North News,* May 15, 2023, accessed June 11, 2023, https://www.highnorthnews.com/en/lavrov-arctic-councils-future-depends-whether-civilized-dialogue-can-continue.
166. Ministry of Foreign Affairs of the Russian Federation, "Foreign Minister Sergey Lavrov's video address to the participants in the 13th session of the Arctic Council, May 11, 2023," accessed June 11, 2023, https://mid.ru/en/foreign_policy/rso/1869388/.
167. "Foreign Minister Sergey Lavrov's video address to the participants in the 13th session of the Arctic Council, May 11, 2023."
168. Interview with a high-ranking official, February 6, 2024.
169. "Russia's future in Arctic Council depends on whether its activity meets Russian interests—Russian Foreign Ministry," *Interfax,* February 5, 2024, accessed February 11, 2024, https://interfax.com/newsroom/top-stories/99094/.
170. "Russia Wants to Cooperate With BRICS Countries on Research on Svalbard," High North News, April 14, 2023, accessed February 10, 2024, https://www.highnorthnews.com/en/russia-wants-cooperate-brics-countries-research-svalbard.
171. "Barents Euro-Arctic cooperation: Joint Statement of the European Union, Finland, Denmark, Iceland, Norway and Sweden on suspending activities with Russia," March 9, 2022, accessed May 29, 2023, https://www.eeas.europa.eu/eeas/barents-euro-arctic-cooperation-joint-statement-european-union-finland-denmark-iceland-norway_en.
172. "Northern Dimension Policy: Joint Statement by the European Union, Iceland and Norway on suspending activities with Russia and Belarus," March 8, 2022, accessed May 29, 2023, https://ndpculture.org/news/northern-dimension-policy-joint-statement-by-the-european-union-iceland-and-norway-on-suspending-activities-with-russia-and-belarus/.
173. "Statement published by the Norwegian MFA: Russia Suspended from the Council of the Baltic Sea States," March 3, 2022, accessed May 29,

2023, https://www.regjeringen.no/en/aktuelt/russland-suspenderes-fra-ostersjoradet/id2903009/. Norway held the presidency of the Council when Russia invaded Ukraine.

174. "Council of the Baltic Sea States Terms of Reference," March 1992, accessed May 29, 2023, https://cbss.org/wp-content/uploads/2020/06/Council-of-the-Baltic-Sea-States-Terms-of-Reference.pdf.
175. "Comment by Foreign Ministry Spokeswoman Maria Zakharova on the situation around the Northern Dimension and the Barents Euro-Arctic Council (BEAC)," March 11, 2022, accessed May 29, 2023, https://mid.ru/ru/foreign_policy/news/1803807/?lang=en.
176. Ministry of Foreign Affairs of the Russian Federation, "Foreign Ministry Statement on the withdrawal of the Russian Federation from the Council of the Baltic Sea States," May 17, 2022, accessed May 29, 2023, https://www.mid.ru/en/foreign_policy/news/1813674/.
177. Ministry of Foreign Affairs of the Russian Federation, "Foreign Ministry Statement on Russia's withdrawal from the Barents Euro-Arctic Council," September 18, 2023, accessed February 2, 2024, https://mid.ru/en/foreign_policy/rso/1904899/.
178. "AEC appoints a new chair from the North Norwegian Company," Arctic Economic Council, May 30, 2023, accessed, May 31, 2023, https://arcticeconomiccouncil.com/news/aec-appoints-a-new-chair-from-the-north-norwegian-company/.
179. Interview with a high-level Icelandic official, April 12, 2023; see also Timo Koivurova and Akiho Shibata, "After Russia's invasion of Ukraine in 2022: Can we still cooperate with Russia in the Arctic," *Polar Record*, 59 (2023) (e12), March 17, 2023, accessed May 29, 2023, 1–9, https://doi.org/10.1017/S0032247423000049.
180. Interview with a high-ranking Icelandic official, April 12, 2023.
181. Interview with a high-ranking Icelandic official, April 12, 2023.
182. See "Putin works with Xi to freeze Nato out of Arctic riches," *The Times*, April 29, 2023, accessed May 1, 2023, https://www.thetimes.co.uk/article/china-and-russia-strike-deal-in-push-for-arctic-powermtd2dpnq8#:~:text=Russia%20and%20China%20have%20carved,meeting%20in%20Murmansk%2C%20northwest%20Russia.
183. Evan T. Bloom, "After a 6-month Arctic Council pause, it's time to seek new paths forward," *Arctic Today*, September 6, 2022, accessed January 24, 2023, https://www.arctictoday.com/after-a-6-month-arctic-council-pause-its-time-to-seek-new-paths-forward/#:~:text=paths%20forward%20%2D%20ArcticToday-,After%20a%206%2Dmonth%20Arctic%20Council%20pause%2C%20it's%20time%20to,for%20collaboration%20in%20the%20region.&text=The%20Arctic%20was%20supposed%20to,pressures%20of%20great%20power%20politics.

184. Oran Young, "Can the Arctic Council Survive the Impact of the Ukraine Crisis?" *Georgetown Journal of International Affairs*, December 30, 2022, accessed January 24, 2023, https://gjia.georgetown.edu/2022/12/30/can-the-arctic-council-survive-the-impact-of-the-ukraine-crisis/.
185. Natalia Drozdiak and Danielle Bochove, "The Arctic Is the New Spot for NATO and Russia to Flex Their Military Muscle," *Bloomberg*, March 13, 2023, accessed February 11, 2024, https://www.bloomberg.com/news/features/2023-03-13/nato-russia-are-flexing-their-military-power-in-the-arctic.
186. "NATO 2022 Strategic Concept," March 3, 2023, accessed February 11, 2024, https://www.nato.int/cps/en/natohq/topics_210907.htm#:~:text=The%202022%20Strategic%20Concept%20describes,and%20management%3B%20and%20cooperative%20security.
187. NATO, "Pre-ministerial press conference by NATO Secretary General Jens Stoltenberg ahead of the meetings of NATO Defence Ministers," October 11, 2022, accessed December 10, 2023, https://www.nato.int/cps/en/natohq/opinions_208037.htm.
188. Icelandic Ministry for Foreign Affairs, "Iceland suspends embassy operations in Moscow," June 9, 2023, accessed June 11, 2023, https://www.government.is/diplomatic-missions/embassy-article/2023/06/09/Iceland-suspends-embassy-operations-in-Moscow/.
189. "Iceland suspends embassy operations in Moscow."
190. "Iceland suspends embassy operations in Moscow."
191. "Russia says Iceland will 'destroy' ties by suspending embassy operations," *Reuters*, June 10, 2023, accessed June 11, 2023, https://www.reuters.com/world/europe/russia-says-iceland-destroys-ties-by-suspending-embassy-operations-2023-06-10/.
192. "Russia says Iceland will 'destroy' ties by suspending embassy operations."
193. "Russia will react to Iceland's unfriendly steps," *Prensa*, June 10, 2023, accessed June 11, 2023, https://www.plenglish.com/news/2023/06/10/russia-will-react-to-icelands-unfriendly-steps/.
194. "Russia will react to Iceland's unfriendly steps."
195. On Iceland's trade relations with the Soviet Union and the East bloc, see, for example, Jón Ólafsson, *Kæru félagar. Íslenskir sósíalistar og Sovétríkin 1920–1960* [Dear Comrades: Icelandic Socialists and the Soviet Union, 1920–1969] (Reykjavík: Mál og menning, 1999); Valur Ingimundarson, *Í eldlínu kalda stríðsins. Samskipti Íslands og Bandaríkjanna 1945–1960*; Skafti Ingimarsson, "Íslenskir kommúnistar og sósíalistar: Flokksstarf, félagsgerð og stjórnmálabarátta 1918–1968" [Icelandic Communists and Socialists: Party Activism, Social Structure, and Political Struggle], Ph.D. diss. (University of Iceland, 2018); Þór Whitehead, "Austurviðskipti

Íslendinga" [Iceland's Trade with the East Bloc], *Frelsið* 3, no. 3 (1982): 198–211.

196. Interview with a high-level governmental official, June 22, 2023.
197. "Sendiráðið í Moskvu verður áfram lokað" [The Embassy in Moscow will Remain Closed], *Morgunblaðið*, October 19, 2023, accessed February 10, 2024, https://www.mbl.is/frettir/innlent/2023/10/19/sendiradid_i_russlandi_verdur_afram_lokad/. Benediktsson repeated practical rationale for the decision: the low level of embassy activities, the idle state of the bilateral relationship, and the need for an ambassadorial rotation. Yet, he stressed Iceland was not severing diplomatic ties with Russia, even if it did not maintain normal embassy functions.
198. Interview with a high-level governmental official, June 22, 2023; "Sendiráðið í Moskvu verður áfram lokað."

CHAPTER 5

Conclusion

Since the end of the Cold War, Iceland's approach toward the Arctic has been driven by two central concerns: that it be counted as a legitimate Arctic state and that its geostrategic location be used to promote political and economic aims. It partly explains the belabored status-seeking attempts to define Iceland's Arcticness as being exceptional or privileged—whether to buttress the questionable claim that it is the only coutry situated as a whole within the Arctic region or to push, until recently, for external recognition of its status as an Arctic coastal state. As we have seen here, this national narrative was met with skepticism by some other Arctic states. To be sure, despite strong Soviet and, later, Russian reservations, Iceland was included in the Arctic institutional arrangements that were formed in the 1990s, notably, the Arctic Council, but also other regional organizations—where its stakeholding role or interests were minimal—such as the Barents Euro-Arctic Council, the Council of the Baltic Sea States, and the Northern Dimension. Yet the Arctic Five were initially not prepared to accept Iceland's coastal state aspirations, even if they belatedly recognized its maritime interests in the Arctic by including it in the negotiations on the fishing moratorium in the central Arctic Ocean.[1]

It has been shown here that Iceland did not pay much attention to the Arctic until in the latter half of the 2000s. It was far more preoccupied with its role in the European integration process—through the EEA and Schengen agreements—and with its failed attempts to maintain the status

© The Author(s), under exclusive license to Springer Nature Switzerland AG 2024

V. Ingimundarson, *Iceland's Arctic Policies and Shifting Geopolitics*, https://doi.org/10.1007/978-3-031-40761-1_5

quo in its defense relationship with the United States after the end of the Cold War. The danger of climate change and global warming was certainly highlighted when Iceland chaired the Arctic Council in the early 2000s. Still, it was not until after the United States closed its military base in Iceland that political elites began to look to the Arctic as a way to open up new hedging possibilities.

Coinciding with broader general geopolitical interest in the Arctic following Russia's North Pole flag-planting, Iceland's focus on the region increased further when it faced the fall-out of the financial crisis. The initial panic-stricken response to the banking collapse—the overture to Russia—did not sit well with Iceland's traditional Western partners. Subsequently, the Arctic became, in a sense, a defensive displacement mechanism of Icelandic elites—after experiencing Western "disavowal"—to redefine Iceland's political identity and to forge a narrative of future potentials. All kinds of economically driven Arctic imaginaries were articulated, such as Iceland's involvement in oil extraction, its future role in transpolar transport, and its hosting of an International Rescue and Response Center. Given the long time span needed for such projects, whose implementation was dependent on the willingness of other states to take part in them, they were as much based on wishful thinking as on economic realities. Needless to say, they could not reverse the downward economic spiral generated by the banking collapse. Still, they gave politicians some breathing space during the financial crisis, especially at a time when public wrath was directed at them for colluding with reckless financial elites and for adopting economic policies that were blamed for the disaster. Thus, the "Arctic turn" reflected a discursive strategy to compensate—geopolitically and economically—for the loss of Iceland's military importance in a post-Cold War setting and to offer a material vision based on the Arctic's future potentials following the national trauma caused by the financial crisis.[2]

Iceland's approach toward the Arctic was also directly influenced by other states and regional institutions. In the 1990s, the other Nordic countries played an important role in ensuring that Iceland became part of the Arctic institutional structure in the face of Russian resistance. Norway's High North policy raised the Icelandic government's awareness of the prospective geopolitical importance of the Arctic, especially after the U.S. military's withdrawal from Iceland. It was, further, underscored by the Icelandic support for Norway's proposal for a NATO monitoring role in the Arctic and for its inclusion in oil exploration in the Dragon Area.

The conclusion of soft security agreements with Norway and Denmark—with limited Swedish and Finnish involvement in military exercises in Iceland—was meant to strengthen the Nordic dimension of Iceland's foreign and security policy. While there was no interest in forming a Nordic bloc in Arctic institutions, Iceland fully agreed with the other Nordic states about granting Asian states, including China, an observer status in the Arctic Council. Iceland was, to be sure, miffed by Danish and Norwegian efforts to facilitate the creation of the Arctic Five and teamed up with the Finns and the Swedes as well as with the Greenlanders in resisting the initiative. Iceland's interest in cultivating ties with Greenland and the Faroe Islands outside the Nordic institutional framework was linked to its role in the West Nordic Council. By being part of a venue that does not include Denmark and Norway, Iceland's status in the region was enhanced, even if the project has failed to develop a distinct policymaking identity of its own in the Arctic.

While Iceland sought to cement its relationships with other Western and Nordic countries, it fully subscribed to the post-Cold War cooperative narrative on the Arctic, with its emphasis on an orderly resolution of regional disputes. True, Iceland's overture to NATO after the closure of the military base was tied to lingering perceptions of Russia as a potential threat, which also explains the irritation with its bomber flights near the island. Yet after the financial crash, the Icelandic government was briefly willing to turn to Russia as a savior. What is more, the plan for the rescue center in Keflavík did not rule out the inclusion of Russia. This position partly made the United States hesitant to endorse the project.[3] It was not, however, until tensions resumed in Western-Russian relations following the annexation of Crimea that it was gradually abandoned.

The push for expanding relations with China was based on the Icelandic government's post-crisis hedging agenda, which combined an EU membership bid with a Free Trade Agreement and Arctic projects with China. As was the case with the Russian loan request prior to and after the financial crisis, Iceland's newfound enthusiasm for China raised suspicions in the West. It started with the Huang Nubo case, which put the international media spotlight on Iceland as a gateway for Chinese Arctic ambitions, and it was followed by probing questions about the bilateral FTA, the size of the Chinese embassy in Reykjavík, the oil exploration collaboration in the Dragon Area, the Finnafjörður port project, and the China-Icelandic Arctic Science Observatory. Some Western claims about the nature of the Sino-Icelandic relationship were highly exaggerated. The

FTA involved no investments and has only had a modest impact on Iceland's foreign trade; Britain and several EU countries as well as the United States continue to be Iceland's core trading partners. That the Chinese should have followed in the footsteps of the Norwegians by pulling out of the Dragon Area project suggests that concerns about its economic viability were more important than potential political benefits. And the Finnafjörður project—symbolizing economic opportunities in connection with the future opening of the transarctic route—has not moved beyond the idea stage. U.S. military interest in harbor facilities in the area could also complicate any such plans, even if Iceland has up to now resisted its militarization.

The government has, in fact, steadily been de-emphasizing material benefits stemming from the Arctic as it has become clear how unrealistic they seem. Yet one can still detect residues of an earlier tendency to "balance" economic and environmental concerns by Iceland's preoccupation—when it chaired the Arctic Council—with deepening the cooperation between the AC as an environmental organization and the Arctic Economic Council with its natural resource utilization agenda.[4] Non-state company interests had been instrumental in pushing the Dragon Area initiative and making it part of the government's Arctic policy, with Eykon Energy taking the lead. The Arctic Science Observatory was also promoted by non-governmental entities, such as the Icelandic Centre for Research and the University of Akureyri, even if municipal interests in North-East Iceland were also important. Hence, the belated and reluctant governmental involvement in the CIAO venture. Legitimate questions have been raised about the financing of the Observatory because it is funded by the Chinese with no direct Icelandic matching contributions. It cannot either be ruled out that the facility could serve dual-use functions. What this case—like that of Huang Nubo—also shows is how local interests driven by economic motives can override foreign policy concerns. Still, given that Icelandic scientists have access to the scientific data and have shared them with Western and Japanese scientists, there is no reason to jump to hasty conclusions about the Observatory's future military implications based on an ideologically fixed "strategic competition" narrative, involving the United States and China, in which there is little or no room for third-party agency or stakes.

In general, there has been a tendency to transpose a 2013 Sino-Icelandic discourse to the present as if nothing happened in the meantime. What needs to be considered are two things: First, Iceland's initial overture to

China was part of an economically determined hedging bid in a time of deep uncertainties after the banking collapse. Western governments were, as discussed here, highly reluctant to come to Iceland's rescue on the grounds that it was, to a large degree, a self-made disaster. The implication was that since the Icelanders had scornfully rejected external warnings about the non-sustainability of the banking expansion abroad, they should be taught a lesson. After closing its military base in Iceland, the United States saw, for example, no reason to serve as a guarantor of Iceland's economic stability as it had done during the Cold War. Second, Iceland's surprisingly quick economic recovery eased the pressure on successive governments to engage in foreign policy experiments outside the Western orbit, for example, by courting China or by pursuing non-traditional economic opportunities in places such as the Arctic.

Yet the 2013 launching of the Arctic Circle Assembly fitted into Iceland's strenuous efforts to promote an Arctic political agenda to underpin a post-crisis material success story. Even if it was not a government project, it quickly became part of Icelandic nation branding. It was rationalized by Ólafur Ragnar Grímsson as a venue to highlight climate change and, far less credulously, to single out Iceland's central role in it. Still, it turned into a personal legacy project targeted at international audiences, where Grímsson reigned supreme under the approving gaze of Icelandic political elites. The failure of some foreign observers to understand the nature of this relationship—with the Icelandic government playing the second fiddle and no direct part in organizing the annual conference—led to wild speculations and suspicions about the motives behind the Arctic Circle, which bordered on the conspiratorial. Interventions that described Grímsson's initiative as a direct challenge to the Arctic Council, as a rebellious government-inspired act against the Arctic Five, or as serving the geopolitical interests of China are cases in point. The same applies to jealous attempts to devalue the Arctic Circle conference—in parochial terms—because it was seen as a competitive threat to another nation branding project: Norway's Arctic Frontiers conference.

How seriously the Arctic Circle was taken—from the start—outside Iceland was, of course, a testament to Grímsson's success in framing the project's misplaced governance ambitions, in using his presidential position to promote it, and, after he left office, in portraying himself as a sort of a global Arctic ambassador and salesman. It also created, however, the false impression that he was responsible for Iceland's Arctic policies and its main enforcer. What partly explains it—as has been stressed here—is that no attempt was made to resolve or negotiate the contradiction between Grímsson's inclusive vision of the global Arctic and the Icelandic government's more exclusive regional one. While the Icelandic government has generally been in favor of broadening the activities of the Arctic Council by supporting observer applications, it has consistently insisted on the privileged role of the Arctic Eight in Arctic governance. Since the last-minute decision to include Grímsson's Arctic Center in the 2021 Arctic policy did not address this tension, it only added to the confusion by reinforcing the projection of the Arctic Circle Assembly as a celebratory national branding venue.

Despite their cross-purposes, Grímsson's Arctic views and those of the Icelandic government were aligned in some other areas. The elitism of Arctic Circle has always been balanced by the promotion of Indigenous interests, with representatives from Greenland and other self-governing, autonomous, or regional entities within states having been provided with a prominent role. This was consistent with Össur Skarphéðinsson's emphasis—during his tenure as Foreign Minister—on expanding relations with Greenland and the Faroe Islands. Later, as head of the Greenland committee, he continued to map out possibilities for increased bilateral cooperation—through the laundry list of proposals—including the presumptuous plan to use the Arctic Circle venture to enhance Greenland's representation and visibility on the world stage. The report's Icelandic-centrist and didactic tone was reflected in its portrayal of Iceland as a model for Greenland's status-seeking efforts, culminating in independence, which were projected in linear and teleological terms. Yet Skarphéðinsson materialist Arctic vision—based on his ideas about making Iceland a future oil-producing state and a service center for natural resource extraction in Greenland—was combined with ideological solidarities with the non-privileged. As Foreign Minister, he had successfully resisted major funding cuts to international development cooperation when Iceland was facing a severe budget squeeze following the banking collapse; as described above, he backed Indigenous rights in Iceland's 2011 Arctic policy and, in 2020, he pushed for support for social, education, and health projects in Greenland.

It is revealing that Grímsson and Skarphéðinsson, who had contributed most to the promotion of Iceland's place in the Arctic in their respective capacities as President and Foreign Minister, should join hands after leaving public office.[5] Skarphéðinsson became the first chair of the board of the Ólafur Ragnar Grímsson Center, whose aim was to cement Grímsson's presidential legacy, before the former President assumed that role himself in 2023. In the beginning, there was no inconsistency between Grímsson's projection of the Arctic as a peaceful area of cooperation—where Great Power rivalries had no place—and Iceland's first Arctic policy as framed by Skarphéðinsson and ratified by the Icelandic parliament. True, Grímsson's consistent opposition to Iceland's EU membership was not welcomed by Skarphéðinsson or the Social Democratic members of the Left-Wing government. His world view and actions, however, were fully in line with Iceland's hedging attempts with respect to China during and after the financial crisis. After Russia's 2014 military intervention in Ukraine, the gap between Grímsson's geopolitical opinions and those of Icelandic government ministers widened.

We have seen that after criticizing the Norwegians openly for using an Arctic venue to attack the Russians for the annexation of Crimea, Grímsson rationalized Putin's invasion of Ukraine as a response to NATO's eastern expansion and was concerned with presenting a "balanced" view of the war, even if he claimed not to support it and lamented the fate of its victims. The rhetoric of Icelandic ministers on Ukraine struck a noticeably different note. Foreign Minister Gunnar Bragi Sveinsson had condemned Russia's annexation of Crimea from the start, and after its 2022 invasion of Ukraine, Foreign Minister Þórdís Kolbrún Reykfjörð Gylfadóttir similarly adopted a hardline anti-Russia line.[6] Yet, like her predecessors—Sveinsson and Þórðarson—Gylfadóttir refrained from publicly criticizing the former President who had been given so much credit for promoting Iceland, internationally, through the Arctic Circle Assembly.

When the Americans renewed their focus on Iceland, it was not only about responding to increased Russian military activities after the Ukraine crisis but also about stemming China's influence. It soon became clear that their interpretation of the Sino-Icelandic economic relationship was outdated. The formal classification of China as the main strategic competitor to the United States after the Trump Administration came to power meant that any Chinese "footprint" in Western countries was treated with suspicion. The positive U.S. reaction to the Icelandic suggestion of concluding a bilateral Free Trade Agreement between the United States and Iceland had no doubt much to do with the latter's FTA with China, even if no decision has been made on the matter. The Trump Administration did not, however, succeed in pressuring Iceland to side decisively with it in its global rivalry with China. Having no military presence in the Arctic or North Atlantic, China was not seen as a threat. Still, while Iceland has not, formally, decided on whether to take part in the Polar Silk Road, its continued silence on the issue—a position consistent with that of other Nordic states—can be interpreted as a clear indication of non-interest and non-inclusion. The Icelandic Foreign Ministry's increased scrutiny of the Observatory project is another case of a more distant and circumspect Icelandic approach toward China. In other words, Iceland's maneuvering space to hedge in its foreign policy—as it did after the financial crash—has ended due to systemic repolarization pressures generated by increased Western-Russian tensions and the U.S.-Sino rivalry. It is reflected in the declining Chinese economic influence in the Arctic, with Arctic Silk and Road projects having been scaled down or scrapped altogether. Among projects that have failed to materialize include a railway connection between northern Finland and Norway; a uranium and rare earth mining site at Kuannersuit in Greenland; a liquefied natural gas investment in Alaska; the Dragon field oil exploration project in Iceland; a land purchase in Svalbard; and an underwater Arctic communications conduit along the Northern Sea Route.[7]

The first signs of Iceland moving back to a balancing or bandwagoning policy could be detected after the annexation of Crimea, when it allowed the United States to resume anti-submarine warfare operations. To be sure, the Ukrainian crisis had a "delayed" effect on Iceland's Arctic policy, which continued to subscribe to an Arctic exceptionalist narrative. Thus, despite his critical statements on Russia, Foreign Minister Gunnar Bragi Sveinsson stuck to the portrayal of the region in cooperative terms. It was also consistent with a broad understanding of the security concept, which was made central in Iceland's 2016 National Security Policy, with its focus on non-military risks. But during Guðlaugur Þór Þórðarson's tenure as Foreign Minister, a traditional North Atlantic military agenda steadily gained ground at the expense of a cooperative Arctic one. The repercussions of the labeling of Russia and China by the United States as "strategic competitors," which included a criticism of their respective roles in the Arctic, were clearly at play here.

The symbolic visits of Vice President Pence and Secretary of State Pompeo confirmed U.S. reengagement with the North, in general, and Iceland, in particular. The Icelandic government was cautious in its dealings with the Trump Administration[8] and was slow to accept and postponed decisions on U.S. requests for additional military projects in Iceland. It reflected the ideological reluctance of the Left Greens to be associated with the remilitarization of Iceland. But Þórðarson became far more direct in questioning Russia's military activities and China's economic influence in the Arctic. This hardening of attitudes was clearly reflected in the explanatory memorandum accompanying Iceland's 2021 revised Arctic policy. Þórðarson justified increased U.S. presence in Iceland by referring to Russia and China. The Americans have, in fact, reassumed a permanent anti-submarine warfare mission role, which they performed in the last phase of the Cold War. That does not mean that the reemergence of the GIUK gap as a strategic choke point in the North Atlantic is comparable to the importance it enjoyed during the East-West conflict. And Russia's invasion of Ukraine has focused NATO's attention far more on the Eastern flank than the North. In strategic terms, however, the return of American military force to Iceland—even if it is still being presented in terms of rotational deployments—signifies the revisiting of a past directed at a hostile Russia.

The revision of Iceland's National Security Policy in 2023 did not lead to any new formal policy chances after the Ukraine War. It suited a three-party government, with ideological differences, which was not prepared to reopen a Cold War debate over a permanent U.S. military presence, while, at the same time, presenting a united front against Russia's assault on Ukraine. Yet the war has tilted Iceland's foreign policy to a clear-cut balancing policy against Russia. As an unarmed state, Iceland has no direct say over Arctic military developments, except for making its territory available to the United States and NATO.[9] With the return of Cold War-style tensions, there are few possibilities for a small state to engage in hedging or to use its place in the Arctic to promote foreign policy experiments as Iceland did for a brief period of time after the financial crisis. As a member of a Western club, Iceland's conformist behavior within it is now reflected in its Arctic policies.

Notes

1. On the agreement, see Cayla Calderwood and Frances Ann Ulmer, "The Central Arctic Ocean fisheries moratorium: A rare example of the precautionary principle in fisheries management," *Polar Record* 59 (2023) (e1), January 16, 2023, accessed May 29, 2023: 1–14, https://doi.org/10.1017/S0032247422000389; Jon Rahbek-Clemmensen and Gry Thomasen. "How has Arctic coastal state cooperation affected the Arctic Council?" Joji Morshita, "The Arctic Five-plus-Five process on central Arctic Ocean fisheries negotiations: Reflecting the interests of Arctic and non-Arctic actors."

2. See Valur Ingimundarson, “Territorial Discourses and Identity Politics: Iceland’s Role in the Arctic.” Ingimundarson, “Iceland’s Post-American Security Policy, Russian Geopolitics and the Arctic Question.”
3. Interview with a high-level U.S. State Department official, June 30, 2014.
4. See also Icelandic Ministry for Foreign Affairs. *Norðurljós. Skýrsla starfshóps um efnahagstækifæri á norðurslóðum.*
5. While characterizing his relationship with Grímsson as that of a “mutual admiration society,” Össur Skarphéðinsson diverged from the former President on Ukraine by being highly critical of Russia’s invasion. See “Pólitískur langhlaupari” [A Political Long-distance Runner], *Morgunblaðið*, July 8, 2023, accessed July 15, 2023, https://www.mbl.is/greinasafn/grein/1839699/?t=287169512&_t=1690133439.100768.
6. Bjarni Benediktsson briefly succeeded Gylfadóttir as Foreign Minister from October 2023 to April 2024. But when he became Prime Minister after Katrín Jakobsdóttir resigned as Prime Minister in the spring of 2024 to run—unsuccessfully—for the position of President of Iceland, Gylfadóttir returned as Foreign Minister.
7. See Marc Lanteigne, “The Rise (and Fall?) of the Polar Silk Road.”
8. U.S. Department of Defense, *Summary of the 2018 National Defense Strategy: Sharpening the American Military Competitive Edge*, 3; see also Department of Defense, *Arctic Strategy* (June 2019), accessed November 30, 2022, https://dod.defense.gov/Portals/1/Documents/pubs/Report_to_Congress_on_Resourcing_the_Artic_Strategy.pdf; United States Coast Guard, *Arctic Strategic Outlook*.
9. See Marzia Scopelliti and Elena Conde Pérez, “Defining security in a changing Arctic: helping to prevent an Arctic security dilemma,” *Polar Record* 52, no. 6 (2016): 672–679.

Bibliography

Archival Material

Þjóðskjalasafn Íslands [ÞÍ] (National Archives of Iceland).

Utanríkisráðuneytið [Icelandic Ministry for Foreign Affairs] 1981, Folder, Jan Mayen, B/9, sendiráð Íslands í Washington [Icelandic Embassy in Washington, D.C.].

Utanríkisráðuneytið 2011a, Folder, Norðurheimssvæði [The Arctic], B/31, 12.P.1, sendiráð Íslands í Osló.

Utanríkisráðuneytið 2011b, Folder, B/530, bréfasafn [Letter Collection], 1941–1994.

Utanríkisráðuneytið 2011c, Folder, B/542, bréfasafn, 1973–1994, Svalbarði [Svalbard/Spitsbergen].

Utanríkisráðuneytið 2011d, Folder, B/549, bréfasafn, 1965–1994.

Utanríkisráðuneytið 2011e, sendiráð Íslands í Brussel [Icelandic Embassy in Brussels], Folder, B/225, 1991–1999.

Utanríkisráðuneytið, sögusafn utanríkisráðuneytisins [Historical Records of the Ministry for Foreign Affairs].

Iðnaðar- og viðskiptaráðuneytið [Ministry of Industry and Commerce], Folder, F/0026, Skýrslur [Reports] 1999–2009.

Iðnaðar- og viðskiptaráðuneytið, Folder, B/1319, B-bréfasafn, 2004–2007.

Iðnaðar-og viðskiptaráðuneytið 2013, Folder, B/807, B-bréfasafn, 1991–1992.

Sjávarútvegsráðuneytið [Ministry of Fisheries], Folder, B/2006, Norðurskautsráðið [Arctic Council], 1996–1997.

Sjávarútvegsráðuneytið 2009, Folder, B/584, 1998–2000.

© The Author(s), under exclusive license to Springer Nature Switzerland AG 2024

V. Ingimundarson, *Iceland's Arctic Policies and Shifting Geopolitics*,
https://doi.org/10.1007/978-3-031-40761-1

Dwight D. Eisenhower Presidential Library, Abilene, Kansas, United States
Dennis Fitzgerald Papers, 1945–69, Box 44.
National Archives, Kew, United Kingdom
Questions on Legal Status Continental Shelf around Spitsbergen, 1974, Foreign Commonwealth Office [FCO] 33/2594.
Legal Status of Spitsbergen, 1976. 59, FCO 33/3071.
Norway and Svalbard (Spitzbergen) FCO 33/4297.
"Public Library of US Diplomacy," WikiLeaks Documents
"The 200-mile Fishery Zone and our Svalbard Policy."

Interviews

Icelandic Foreign Ministry official, interview by the author, August 28, 2004.
High-ranking Icelandic official, interview by author, November 15, 2008.
High-ranking Icelandic official, interview by author, January 29, 2008.
Foreign Ambassador posted to Iceland, interview by author, June 18, 2009.
High-level Russian official, interview by author, June 22, 2009.
U.S. defense officials, interviews by author, September 23, 2009.
Former Icelandic minister, interview by author, November 9, 13, 28, 2011.
Senior Arctic official, interview by author December 3, 2012.
Permanent Representatives to NATO and Senior NATO officials, interviews by author, December 10 and 12, 2012.
Senior Arctic officials, interviews by author, December 12, 2012.
High-level Icelandic official, interview by author, December 20, 2012.
Icelandic government minister, interview by author, April 10, 2013.
Senior Arctic officials, interviews by author, May 18, 2013a.
Senior Arctic officials, interviews by author, May 28, 2013b.
High-ranking Icelandic defense official, interview by author, November 11, 2013.
High-ranking Icelandic defense officials, interview by author, June 20, 2014.
High-ranking U.S. State Department official, interview by author, June 30, 2014.
Former Icelandic government minister, interview by author, July 2, 2014.
Canadian official, interview by author, December 3, 2014.
High-ranking Icelandic Arctic official, interview by author, January 20, 2017a.
High-ranking Icelandic defense official, interview by author, March 29, 2017b.
U.S. Department of Defense official, interview by author, April 10, 2018.
High-ranking Icelandic security official, interview by author, January 24, 2019a.
High-ranking Icelandic government official, interview by author, February 5, 2019b.
Icelandic security official, interview by author, February 5, 2019.
High-ranking Icelandic security officials, interview by author, November 15, 2019c.
High-ranking Icelandic Arctic official, interview by author, March 12, 2021a.

High-ranking Icelandic security official, interview by author, May 11, 2021b.
High-ranking Icelandic security official, interview by author, April 23, 2021c.
High-ranking Icelandic security official, interview by author, September 7, 2021d.
High-ranking Icelandic security officials, interview by author, February 5, 2021e.
High-ranking Icelandic Arctic official, interview by author, March 5, 2021f.
High-ranking Icelandic Arctic official, interview by author, March 12, 2021.
Icelandic government minister, interview by author, March 7, 2022a.
Icelandic government minister, interview by author, September 9, 2022b.
Icelandic science official, interview by author, December 16, 2022.
High-ranking Icelandic foreign ministry official, interview by author, January 13, 2023a.
Icelandic official, interview by author, January 15, 2023.
High-ranking Icelandic official, interview by author, April 12, 2023b.
Former Senior Arctic Official, July 18, 2022.
High-ranking governmental official, interview by author, June 22, 2023c.
High-ranking Arctic official, interview by author, February 6, 2024.
Icelandic academic, interview by author, February 3, 2024.
Gunnlaugur Björnsson, email communications to author, February 8 and 10, 2024.
Former high-level Arctic Council official, interview with author, February 19, 2024.
Former high-level Icelandic official, interview with author, February 19, 2024.
High-level Icelandic defense official, interview by author, February 23, 2024.
Icelandic official, interview by author, February 23, 2024.
Icelandic official, interview by author, October 20, 2024.

Newspapers and Other Media Outlets

Arctic Institute, Center for Circumpolar Security Studies. "Iceland for Sale – Chinese Tycoon Seeks to Purchase 300 km² of Wilderness," September 2, 2011. https://www.thearcticinstitute.org/2011/09/iceland-for-sale-chinese-tycoon-seeks.html.

"The Arctic contest heats up. What is Russia up to in the seas above Europe." *The Economist*, October 9, 2008.

Askja Energy. The Independent Icelandic and Northern Energy Portal. "Foreign Investment [in Iceland]." n.d. https://askjaenergy.com/iceland-investing/foreign-investment/.

"ASÍ mótmælir fríverslunarsamningi við Kína" [The Icelandic Federation of Labor Protests the Free Trade Agreement with China]. *Vísir*, April 12, 2013. https://www.visir.is/asi-motmaelir-friverslunarsamningi-vid-kina/article/2013130419728.

Belton, Catherine, and Tom Braithwaite. "Iceland asks Russia for €4bn loan after west refuses to help." *Financial Times*, October 8, 2008.

Bennett, Mia. "Arctic Circle 2022: A NATO admiral, Chinese diplomat, and Faroese metal band walk into a concert hall," October 19, 2022. *Cryptopolitics* (blog). https://www.cryopolitics.com/2022/10/19/arctic-circle-2022/.

"Bjarni hafði efasemdir frá upphafi" [Bjarni Benediktsson Harbored Doubts from the Start]. *Morgunblaðið*, August 20, 2015. https://www.mbl.is/frettir/innlent/2015/08/20/bjarni_hafdi_efasemdir_fra_upphafi

Bloom, Evan T. "After a 6-month Arctic Council pause, it's time to seek new paths forward." *Arctic Today*, September 6, 2022. https://www.arctictoday.com/after-a-6-month-arctic-council-pause-its-time-to-seek-new-pathsforward/#:~:text=paths%20forward%20%2D%20ArcticToday,After%20a%206%2Dmonth%20Arctic%20Council%20pause%2C%20it's%20time%20to,for%20collaboration%20in%20the%20region.&text=The%20Arctic%20was%20supposed%20to,pressures%20of%20great%20power%20politics.

Bremensport. "Welcome to the Finnafjord Project." https://bremen-ports.de/finnafjord/.

"Bremenport rannsakar aðstæður í Finnafirði" [Bremenport Explores the Conditions of Finnafjörður]. *Morgunblaðið*, August 30, 2013. https://www.mbl.is/vidskipti/frettir/2013/08/30/bremenports_rannsakar_adstaedur_i_finnafirdi/.

"Bremenports to Develop New Deep-Water Port in Iceland." *World Maritime News*, May 27, 2015. https://worldmaritimenews.com/archives/275015/bremenports-to-develop-new-deep-water-port-in-iceland/.

"British tycoon buys vast farm in the highlands. Intends to do 'absolutely nothing with it.'" *Iceland Magazine*, December 16, 2016. https://icelandmag.is/article/british-billionaire-buys-vast-farm-highlands-intends-do-absolutely-nothing-it.

"Bæta í samstarf Grænlands og Íslands" [The Expansion of Greenlandic-Icelandic Relations]. *Morgunblaðið*, September 23, 2023. https://www.mbl.is/frettir/innlent/2021/09/23/baeta_i_samstarf_graenlands_og_islands/.

Canada rejects Arctic flag-planting as 'just a show by Russia'." *The Sidney Morning Herald*, August 3, 2007. https://www.smh.com.au/world/canada-rejects-arctic-flagplanting-as-just-a-show-by-russia-20070803-r86.html.

Capell, Kerry. "The Stunning Collapse of Iceland." *NBC*, October 9, 2008. https://www.nbcnews.com/id/wbna27104617.

"China Begins to Revive Arctic Scientific Ground Projects After Setbacks." *Voice of America*, December 5, 2022. https://www.voanews.com/a/china-begins-to-revive-arctic-scientific-ground-projects-after-setbacks-/6860756.html#:~:text=Beijing%20is%20taking%20its%20first,%22near%2DArctic%22%20state.

"Chinese Container Ship Transits Arctic, More Oil Tankers and Massive Bulk Carrier Also En Route." *High North New*, August 21, 2023. https://www.

highnorthnews.com/en/chinese-container-ship-transits-arctic-more-oil-tankers-and-massive-bulk-carrier-also-en-route.

"Clinton til varnar Íslandi" [[Hillary] Clinton Defends Iceland]. *RUV*, March 30, 2010. https://www.ruv.is/frettir/innlent/clinton-til-varnar-islandi.

"Clinton rebuke overshadows Canada's Arctic meeting." *Reuters*, March 30, 2010. https://www.reuters.com/articl/idUSTRE62S4ZP20100330.

"CNOOC to lead offshore Iceland license group: Orkustofnun, Iceland's National Energy Authority has granted its third license in the offshore Dreki area." *Offshore*, February 6, 2014. https://www.offshore-mag.com/regional-reports/article/16783755/cnooc-to-lead-offshore-iceland-license-group

"Davíð Oddsson: 'Öll aðstoð frá Rússlandi er vel þegin'" [All Assistance from Russia Welcome]. *Viðskiptablaðið*, September 6, 2010. https://www.vb.is/frettir/davi-oddsson-oll-asto-fra-russlandi-er-vel-egin/.

"Drekasvæðið mun betra en við þorðum að vona" [The Dragon Area Has Far More Potential Than We Had Dared to Hope]. *Vísir*, October 11, 2016. https://www.visir.is/g/2016161019705/-drekasvaedid-mun-betra-en-vid-thordum-ad-vona-.

Drozdiak, Natalia. and Danielle Bochove. "The Arctic Is the New Spot for NATO and Russia to Flex Their Military Muscle." *Bloomberg*, March 13, 2023. https://www.bloomberg.com/news/features/2023-03-13/nato-russia-are-flexing-their-military-power-in-the-arctic.

"Ekki mistök að loka varnarstöðinni" [Not a Mistake to Close the Defense Base]. *Morgunblaðið*, April 20, 2023. https://www.mbl.is/frettir/innlent/2023/04/14/ekki_mistok_ad_loka_varnarstodinni/.

"Estonia Says Chinese Ship is Main Focus of Probe Into Cables Damage." *Reuters*, November 10, 2023. https://gcaptain.com/estonia-eyes-chinese-ship-cable-damage-probe.

"'Everything indicates' Chinese ship damaged Baltic pipeline on purpose, Finland says." *Politico*, December 1, 2023. https://www.politico.eu/article/balticconnector-damage-likely-to-be-intentional-finnish-minister-says-china-estonia/.

"Eyjólfur Konráð Jónsson, 13. júní, 1928 – 6. Marz 1997." *Morgunblaðið*, March, 14, 1997. https://timarit.is/page/1874754#page/n0/mode/2up.

"Fagna sögulegum tímamótum með brottför hersins" [Celebrate the Military Withdrawal as a Historical Moment]. *Morgunblaðið*, September 29, 2006. https://www.mbl.is/greinasafn/grein/1105325/?t=926461552&_t=1706893262.5667615.

"Finland says vessel's broken anchor caused Balticconnector damage." *Aljazeera*, October 25, 2023. https://www.aljazeera.com/news/2023/10/25/finland-pipeline-damage-cause.

"First Chinese ship crosses Arctic Ocean amid record melt." *Reuters*, August 17, 2012. https://www.reuters.com/article/china-environment-idUSL6E8JH9AY20120817.

Forsby, Andreas Bøje. "Falling out of Favor: How China Lost the Nordic Countries." *The Diplomat*, June 24, 2022. https://thediplomat.com/2022/06/falling-out-of-favor-how-china-lost-the-nordic-countries/.

"Forsetinn setti ofan í við norskan ráðherra" [The Icelandic President Scolded a Norwegian Minister]." *Vísir*, March 19, 2014. https://www.visir.is/g/2014140318637/forsetinn-setti-ofan-i-vid-adstodarutanrikisradherra-noregs.

"Fyrirhugað að hafa herskip í Helguvík" [The Docking of Warships Planned in Helguvík]. *Morgunblaðið*, June 9, 2023. https://www.mbl.is/mogginn/bladid/grein/1837704/.

"Föst viðvera herliðs á Íslandi hafi fælingarmátt" [A Permanent Military Presence in Iceland Serves a Deterrence Function]. *Morgunblaðið*, March 6, 2022. https://www.mbl.is/frettir/innlent/2022/03/06/fost_vidvera_herlids_a_islandi_hafi_faelingarmatt/.

"Gagnrýnin byggir á vanþekkingu" [The Criticism is Based on Ignorance]. *Morgunblaðið*, January 24, 2023. https://www.mbl.is/mogginn/bladid/grein/1827832/?t=838151989&_t=1674731166.3241313.

"Gagnrýnir áform í Finnafirði" [Criticizes Finnafjörður Plans]. *RUV* [Icelandic Public Radio Broadcasting Services], September 9, 2013. https://www.ruv.is/frett/gagnrynir-aform-i-finnafirdi.

"Geir lýsti yfir óánægju með rússaflug" [PM Geir [Haarde] Expressed Displeasure of Russian Flights]. *Vísir*. n.d. https://www.visir.is/g/200880405055/geir-lysti-yfir-oanaegju-med-russaflug.

"'Greenland is Not for Sale': Trump's Talk of a Purchase Draws Derision." *New York Times*, August 16, 2019. https://www.nytimes.com/2019/08/16/world/europe/trump-greenland.html.

"Halldór Ásgrímsson: Sjónarmið forseta og ríkisstjórnar fari saman" [The President and the Government Should Speak with One Voice]. *Morgunblaðið*, November 25, 1998. https://www.mbl.is/greinasafn/grein/433648/.

Haraldsson, Jónas. "Olíuvinnslan vart áhættunnar virði" [Oil Extraction Hardly Worth the Risk]. *Fréttatíminn*, November 7–9, 2014. https://timarit.is/page/6194027#page/n12/mode/2up.

"Harpa er martröð skattgreiðenda" [Harpa Is the Nightmare of Taxpayers]. RUV, August 4, 2012. https://www.ruv.is/frettir/innlent/harpa-er-martrod-skattgreidenda.

Hellström, Jerker. "China's Political Priorities in the Nordic Countries: from technology to core interests." Norwegian Institute of International Affairs (NUPI), *Policy Brief*, 12 (2016). https://nupi.brage.unit.no/nupi-xmlui/handle/11250/2387468.

"Heræfing NATO á Íslandi." *RUV*, February 1, 2014. https://www.ruv.is/frettir/innlent/heraefing-nato-a-islandi.

"Herstöðvaandstæðingar fagna brottför hersins" [Base Opponents Celebrate the Withdrawal of the [U.S.] Military]. *Vísir*, October 1, 2006. https://www.visir.is/g/20061839069d.

Humpert, Malte. "Chinese Container Ship Completes First Round Trip Voyage Across Arctic." *High North News*, October 9, 2023. https://www.highnorthnews.com/en/chinese-container-ship-completes-first-round-trip-voyage-across-arctic.

"Hætta við kaup á Grímsstöðum" [Abandon Plans for the Purchase of Grímsstaðir]. *RUV*, December 12, 2014. https://www.ruv.is/frett/haetta-vid-kaup-a-grimsstodum.

"Iceland and Germany plan new port." *Maritime Journal*, July 18, 2013. https://www.maritimejournal.com/news101/marine-civils/port,-harbour-and-marine-construction/iceland-plans-new-port.

"Iceland invites China to Arctic Shipping." *The Barents Sea Observer*, September 22, 2010. https://barentsobserver.com/en/sections/business/iceland-invites-china-arctic-shipping.

"Iceland protests five-nation fishing deal in Arctic." *News24*, July 24, 2015. https://www.news24.com/news24/green/news/iceland-protests-five-nation-fishing-deal-in-arctic-20150724.

"Iceland seeks loan from Russia." *Bloomberg*, October 7, 2023. https://www.business-standard.com/article/finance/iceland-seeks-loan-from-russia-108100801025_1.html.

"Icelandic President Ólafur Ragnar Grímsson speaks at April 17, 2013 National Press Club Luncheon." *YouTube*, April 17, 2013. https://www.youtube.com/watch?v=wW0p_Eh94PI.

"Iceland upset by Arctic summit snub." *CBC* News, February 16, 2010. https://www.cbc.ca/news/canada/north/iceland-upset-by-arctic-summit-snub-1.885441.

"Inviterar Russland" [Invites Russia]. *Klassekampen*, November 12, 2008. https://klassekampen.no/utgave/2008-11-12/inviterer-russland.

"Ísland fest sig í sessi sem miðstöð norðurslóða" [The Consolidation of Iceland as a Central Arctic Hub], *Fréttablaðið*, May 21, 2021. https://www.frettabladid.is/frettir/olafur-island-fest-sig-i-sessi-sem-midstod-nordursloda/.

"Ísland í betri stöðu utan ESB" [Iceland in a Better Position Outside the EU]. *Morgunblaðið*, December 23, 2012. https://www.mbl.is/frettir/innlent/2012/12/13/island_i_betri_stodu_utan_esb/.

"Íslensk stjórnvöld hafa ekkert eftirlit eða aðkomu að rannsóknarmiðstöð Kína" [The Icelandic Government Does Not Monitor and Has No Involvement in China's Research Center]. *Heimildin*, June 9, 2023. https://heimildin.

is/grein/18033/islensk-stjornvold-hafa-ekkert-eftirlit-eda-adkomu-ad-rannsoknarmistod-kina/.

"Íslenskur her ekki tímabær" [An Icelandic Military Not Timely]. *Morgunblaðið*, March 7, 2023. https://www.mbl.is/frettir/innlent/2023/03/07/islenskur_her_ekki_timabaer/.

Jónasson, Ögmundur. "Norræn þögn sama og norrænt samþykki" [Nordic Silence Equals Nordic Approval], November 18, 2009. https://www.ogmundur.is/is/greinar/norraen-thogn-sama-og-norraent-samthykki.

Jennings, Gareth. "Nordic countries combine combat air power." *Janes*, March 24, 2023. https://www.janes.com/defence-news/news-detail/nordic-countries-combine-combat-air-power.

Jónasson, Kári. "Árangur af Kínaheimsókn forsetans." *Vísir*, May 20, 2006. https://varnish-7.visir.is/g/2005505200301/arangur-af-kinaheimsokn-forsetans.

"Katrín gefur kost á sér" [Katrín Jakobsdóttir runs for President], *Vísir*, April 5, 2024. https://www.visir.is/g/20242552483d/katrin-gefur-kost-a-ser. The three-party government fell apart in October 2024.

"Katrín segir að Hringborð norðurslóða hafa breytt umræðunni" [PM Katrín Jakobsdóttir Says That the Arctic Circle Has Changed the Debate]. *Vísir*, October 19, 2018. https://www.visir.is/g/2018181018681/f/f/skodanir.

"Katrín segir nýja herstöð ekki koma til greina" [PM Katrín Jakobsdóttir Says That a New Military Base Is out of the Question]. *Vísir*, November 3, 2022. https://185.21.17.249/g/20202032771d/katrin-segir-nyja-herstod-ekki-koma-til-greina.

"Kissinger Phone Call to Fischer Disclosed." *New York Times*, 18 July 1972. https://www.nytimes.com/1972/07/18/archives/kissinger-phone-call-to-fischer-disclosed.html.

Kosningasaga. Upplýsingar um kosningar á Íslandi [History of Elections: Electoral Information in Iceland]. "Forsetakosningar 2012" [Presidential Elections]. https://kosningasaga.wordpress.com/forsetakosningar/forsetakosningar-2012/.

Koivurova, Timo. "The Arctic Council can continue without Russia." *Arctic Today*, March 10, 2022. https://www.arctictoday.com/the-arctic-council-can-continue-without-russia/.

"Lavrov: 'The Arctic Council's Future Depends on Whether a Civilized Dialogue Can Continue.'" *High North News*, May 15, 2023. https://www.highnorthnews.com/en/lavrov-arctic-councils-future-depends-whether-civilized-dialogue-can-continue.

Mastro, Oriana S. "How China is bending the rules in the South China Sea." *The Interpreter*, February 17, 2021. https://www.lowyinstitute.org/the-interpreter/how-china-bending-rules-south-china-sea.

McVicater, Daniel. “How the Russia-Ukraine War Challenges Arctic Governance.” *Council of Foreign Relations*, May 10, 2022. https://www.cfr.org/blog/how-russia-ukraine-war-challenges-arctic-governance.

“Merkilegt skref í samstarfi Íslands og Grænlands” [A Noteworthy Step in Icelandic-Greenlandic Cooperation]. *Fréttablaðið*, October 13, 2022. https://www.frettabladid.is/frettir/merkilegt-skref-i-samstarfi-islands-og-graenlands/.

“Mikil umsvif og náin samvinna” [Extensive [Military] Activities and Close Cooperation].” *Morgunblaðið*, April 13, 2023. https://www.mbl.is/mogginn/bladid/grein/1833634/?t=380088780&_t=1681913551.4705353.

“Mikilvægasta verkefnið” [The Most Important Task], *Þjóðviljinn*, December 30, 1956.

“Mistök að styðja olíuleit á Drekasvæðinu” [A Mistake to Support Oil Exploration in the Dragon Area]. *RUV*, March 22, 2015. https://www.ruv.is/frett/mistok-ad-stydja-oliuleit-a-drekasvaedi.

“Newnew Shipping Line launches regular container service between China and Arkhangelsk.” *PortNews*, July 7, 2023. https://en.portnews.ru/news/350079/.

“No Joke: Trump Really Wants to Buy Greenland.” *National Public Radio* (*NPR*), August 19, 2019. https://www.npr.org/2019/08/19/752274659/no-joke-trump-really-does-want-to-buy-greenland.

“Obama and Nordic Nations Discuss Russia.” *New York Times*, May 13, 2016. https://www.nytimes.com/video/us/politics/100000004405045/obama-and-nordic-nations-discuss-russia.html, May 13, 2016.

“Obama Warms Up to Nordic Leaders.” *New York Times*, May 13, 2016. https://www.nytimes.com/2016/05/14/world/europe/obama-warms-to-nordic-heads-of-state.html.

“Oil exploration in Icelandic waters comes to an end: Too expensive and too risky.” *Iceland Magazine*, January 23, 2018. https://icelandmag.is/article/oil-exploration-icelandic-waters-comes-end-too-expensive-and-too-risky.

“Ólafur Ragnar deilir viðtölum við Rússa um stríðið” [Ólafur Ragnar Grímsson Circulates Interviews with Russians about the War]. *Fréttablaðið,* April 5, 2022. https://www.frettabladid.is/frettir/olafur-ragnar-gagnryndur-fyrir-ad-deila-skodunum-russa-um-stridid/.

“Ólafur Ragnar kominn til Kína.” *Vísir*, May 16, 2005. https://www.visir.is/g/20051182278d.

“Olíuleit í uppnámi” [Oil Exploration in Limbo]. *Viðskiptablaðið*, January 25, 2018. https://www.vb.is/frettir/oliuleit-i-uppnami/.

Piirsalu, Januus. “Russian warplanes cannot switch on transponders.” *Postimees*, September 6, 2016. https://news.postimees.ee/3826371/russian-warplanes-cannot-switch-on-transponders.

"Pólitískur langhlaupari" [A Political Long-distance Runner]. *Morgunblaðið*, July 8, 2023. https://www.mbl.is/greinasafn/grein/1839699/?t=287169512&_t=1690133439.100768.

"Pompeo – Russia is 'aggressive' in the Arctic, China's work there also needs watching." *Reuters,* May 6, 2019. https://www.reuters.com/article/us-finland-arctic-council/pompeo-russia-is-aggressive-in-arctic-chinas-work-there-also-needs-watching-idUSKCN1SC1AY.

"Putin works with Xi to freeze Nato out of Arctic riches." *The Times*, April 29, 2023. https://www.thetimes.co.uk/article/china-and-russia-strike-deal-in-push-for-arctic-power-mtd2dpnq8#:~:text=Russia%20and%20China%20have%20carved,meeting%20in%20Murmansk%2C%20northwest%20Russia.

Rocard, Michel. Address at Arctic Circle Assembly, *Youtube*, November 1, 2014. https://www.youtube.com/watch?v=qleb5wprlhw&list=PLI0a77tmNMvRkshNIRBriGV-K8o3heehr&index=24.

"Russia plants flag on North Pole seabed." *Guardian*, August 2, 2007. https://www.theguardian.com/world/2007/aug/02/russia.arctic.

"Russia hands over presidency of Arctic Council's Scientific Work to Norway." *TASS*, April 14, 2023, https://tass.com/russia/1604697.

"Russia says Iceland will 'destroy' ties by suspending embassy operations." *Reuters*, June 10, 2023. https://www.reuters.com/world/europe/russia-says-iceland-destroys-ties-by-suspending-embassy-operations-2023-06-10/.

"Russia will react to Iceland's unfriendly steps." *Prensa*, June 10, 2023. https://www.plenglish.com/news/2023/06/10/russia-will-react-to-icelands-unfriendly-steps/.

"Russia's Arctic oil exports surge but risks still hamper new trade route." *S&P Global Commodity Insights*, November 29, 2023. https://www.spglobal.com/commodityinsights/en/market-insights/latest-news/oil/112923-russias-arctic-oil-exports-surge-but-risks-still-hamper-new-trade-route#:~:text=Flows%20from%20Russia's%20Arctic%20and,the%20first%20commercial%20shipments%20began.

"Russia's future in Arctic Council depends on whether its activity meets Russian interests – Russian Foreign Ministry." *Interfax,* February 5, 2024. https://interfax.com/newsroom/top-stories/99094/.

"Russian Firm Says Baltic Telecoms Cable was Severed as Chinese Ship Passed Over." *Marine Link*, November 7, 2023. https://www.marinelink.com/news/russian-firm-says-baltic-telecoms-cable-509285.

"Ræddi meðal annars mannréttindamál við forseta Kína" [Discussed Human Rights Among Other Things with the President of China]. *Morgunblaðið*, May 17, 2005. https://www.mbl.is/frettir/innlent/2005/05/17/raeddi_medal_annars_mannrettindamal_vid_forseta_kin/.

Schonberg, Harold C. "Cold War in the World of Chess." *New York Times*, September 27, 1981. https://www.nytimes.com/1981/09/27/magazine/cold-war-in-the-world-of-chess.html.

Schreiber, Melody. "A new China-Iceland Arctic science observatory is already expanding its focus." *Arctic Today*, October 31, 2018. https://www.arctictoday.com/new-china-iceland-arctic-science-observatory-already-expanding-focus/.

"Send verður inn umsókn um aðild að ESB" [An EU Application will be Submitted). *Morgunblaðið*, July 16, 2009. https://www.mbl.is/frettir/innlent/2009/07/16/samthykkt_ad_senda_inn_umsokn/.

"Sigmundur: Mikið í húfi" [PM Sigmundur Davíð Gunnlaugsson: Much at Stake]. *Morgunblaðið*, October 31, 2014. https://www.mbl.is/frettir/innlent/2014/10/31/sigmundur_mikid_i_hufi

"Sigmundur Davíð vill endurmeta EES-samninginn og ekki taka þátt í refsiaðgerðum 'blindandi'" [Sigmundur Davíð Gunnlaugsson Wants to Reevaluate the EEA-Agreement and Is Against "Blindingly" Taking Part in Economic Sanctions]. *Kjarninn*, January 6, 2016. https://kjarninn.is/frettir/2016-01-06-sigmundur-david-vill-endurmeta-ees-samninginn-og-ekki-taka-thatt-i-refsiadgerdum-blindandi/.

Skarphéðinsson, Össur. "Viðbragðsmiðstöð á norðurslóð" [A Rescue Center in the North]. *Morgunblaðið*, March 14, 2013a. https://www.mbl.is/greinasafn/grein/1458610/.

Skarphéðinsson, Össur. "Auknar og breyttar kröfur á forseta" [Additional and Changed Demands on the President]. *Fréttablaðið*, March 23, 2016. https://timarit.is/page/6509020#page/n19/mode/2up.

"Skoða ætti endurkomu hersins" [A Return of the [U.S.] Military Should Be Explored]. *Morgunblaðið*, April 7, 2023. https://www.mbl.is/mogginn/bladid/netgreinar/2023/04/07/skoda_aetti_endurkomu_hersins/.

"Skref í átt að enn betri samvinnu: Fyrsti sendimaður Grænlands á Íslandi tekinn til starfa" [A Step Towards Deeper Cooperation: The First Greenlandic Representative Has Assumed His Functions]." *Morgunblaðið*, January 3, 2019. https://www.mbl.is/greinasafn/grein/1709573/.

Staalesen, Atle. "Newnew Polar Bear sails towards Bering Straits." *The Barents Observer*, November 6, 2023. https://thebarentsobserver.com/en/security/2023/11/newnew-polar-bear-exits-northern-sea-route.

Staalesen, Atle. "Chinese shippers shun Russian Arctic waters," *The Barents Observers*, August 22, 2022. https://thebarentsobserver.com/en/industry-and-energy/2022/08/chinese-shippers-shun-russian-arctic-waters.

"Staðfest að fjölmiðlalögin verði afturkölluð" [It Has Been Confirmed That the Media Law Will Be Withdrawn]. *Morgunblaðið*, July 20, 2004. https://www.mbl.is/frettir/innlent/2004/07/20/stadfest_ad_fjolmidlalog_verda_afturkollud/.

Stavrividis, James. "U.S. Aims at Russia With Navy's Resurrected Second Fleet." *Bloomberg*, May 16, 2018. https://gcaptain.com/u-s-aims-at-russia-with-navys-resurrected-second-fleet-stavridis/.

"Stjórn Húsavíkurstofu styður áform Huang Nubo" [The Tourist Board of the Húsavík Municality Support Huang Nubo's Plans], *640.is*, November 22, 2011. https://www.640.is/is/frettir/stjorn-husavikurstofu-stydur-aform-huang-nubo.

"Stærsta verkefni Íslandssögunnar. Hvað er að gerast í Finnafirði?" [The Greatest Project in the History of Iceland. What Is Happening in Finnafjörður?]. *Kjarninn*, March 21, 2021. https://kjarninn.is/skyring/staersta-verkefni-islandssogunnar-hvad-er-ad-gerast-i-finnafirdi/.

"Støre frykter Norge kan miste nordområdeinitiativet" [Støre Fears That Norway Can Lose the Initiative on the Arctic]. *NRK*, November 10, 2015. https://www.nrk.no/tromsogfinnmark/store-frykter-norge-kan-miste-nordomradeinitiativet-1.12646936.

"Teeing Off at Edge of the Arctic? A Chinese Plan Baffles Iceland." *New York Times*, March 22, 2013. https://www.nytimes.com/2013/03/23/world/europe/iceland-baffled-by-chinese-plan-for-golf-resort.html?pagewanted=all&_r=0.

"Top US, Russia diplomats spar firmly but politely in Iceland." *AP*, May 20, 2021. https://apnews.com/article/donald-trump-iceland-europe-russia-middle-east-24e823ef46207bfbe4c841e1b4a864f0.

"Trú á olíufund myndi strax bæta lánstraust Íslendinga" [A Belief in an Oil Find Would Immediately Improve Iceland's Credit Rating]. *Fréttablaðið*, November 7, 2013. https://timarit.is/page/6142786#page/n0/mode/2up.

"UAE's COP28 president will keep role as head of national oil company." *Guardian*, January 13, 2023. https://www.theguardian.com/environment/2023/jan/13/uae-cop28-president-sultan-al-jaber-to-keep-role-as-head-of-national-oil-company.

Valgreen, Carsten et al. *Iceland: Geyser crisis.* Copenhagen: Danske Bank, 2006. https://www.mbl.is/media/98/398.pdf.

Vasiliev, Anton. "A Story of an Image: The Arctic Council at 25 – Reflections, Arctic Circle." Arctic Circle, September 20, 2021. https://www.arcticcircle.org/journal/by-anton-vasiliev-russias-senior-arctic-official.

Vilhjálmsson, Ingi Freyr. "Þingmaður spyr Katrínu um eftirlit með kínversku rannsóknarmiðstöðinni" [A Parliamentarian Asks [PM Katrín [Jakobsdóttir] about how the Chinese Research Center Is Monitored]. *Heimildin*, April 3, 2023a. https://heimildin.is/grein/17311/.

Vilhjálmsson, Ingi Freyr. "Önnur ríki hafa áhyggjur af norðurljósamiðstöð Kína á Íslandi" [Other Countries Are Worried about the Aurora Observatory]. *Heimildin*, April 1, 2023b. https://heimildin.is/grein/17281/.

Vilhjálmsson, Ingi Freyr. "NATO hefur lýst áhyggjum af rannsóknamiðstöð Kína um norðurljósin" [NATO Has Expressed Worries about China's Northern Lights Research Center]. *Heimildin*, March 28, 2023c. https://heimildin.is/grein/17192/.

Vilhjálmsson, Ingi Freyr. "Kínverska ríkið setur 700 til 800 milljónir í rannsóknarmiðstöð um norðurljósin" [China Invests 700–800 Million [Icelandic Kronor] in an Aurora Observatory]. *Heimildin*, March 6, 2023d. https://heimildin.is/grein/16852/.

"Vill liðka fyrir bættum samskiptum" [Wants to Facilitate Improved Relations]. *Fréttablaðið*, April 16, 2021. https://frettabladid.overcastcdn.com/documents/210416.pdf.

"Voru 153 daga við kafbátaeftirlit á Íslandi [Conducted Submarine Surveillance for 153 Days]," R*ÚV*, November 14, 2018. https://www.ruv.is/frettir/innlent/voru-153-daga-vid-kafbataeftirlit-a-islandi.

Webb, Robert. "Iceland president sounds climate alarm demanding global attention, action at NPC Luncheon." National Press Club, April 15, 2015. https://www.press.org/newsroom/iceland-president-sounds-climate-alarm-demanding-global-attention-action-npc-luncheon.

"Why the Arctic Economic Council is Needed: The People of the Arctic Deserve Our Full Attention and Opportunity to Develop." *High North News*, May 31, 2023. https://www.highnorthnews.com/en/people-arctic-deserve-our-full-attention-and-opportunity-develop.

Xavier-Bender, Guillaume. "China and the West May Soon Compete for Troubled Iceland," May 7, 2013. *RealClear World*. https://www.realclearworld.com/articles/2013/05/07/china_and_the_west_may_soon_compete_for_troubled_iceland_105141.html.

Zarakhovic, Yuri. "Why Russia Is Bailing Out Iceland." *Time*, October 13, 2008. https://content.time.com/time/world/article/0,8599,1849705,00.html.

"Ögmundur segir sölu Grímsstaða mark um vesaldóm stjórnvalda" [Ögmundur [Jónasson] Characterizes the Sale of the Land Grímsstaðir as an Example of the Government's Spinelessness]." *Vísir*, December 20, 2016. https://www.visir.is/g/2016161229918/ogmundur-segir-solu-grimsstada-mark-um-vesaldom-stjornvalda.

"Össur gagnrýnir Ólaf fyrir ummæli um ESB" [Össur Skarphéðinsson Criticizes Ólafur Ragnar Grímsson for His Comments on the EU]. *RUV*, June 6, 2013, https://www.ruv.is/frettir/innlent/ossur-gagnrynir-olaf-fyrir-ummaeli-um-esb.

"Þurfa að skoða nýjar aðferðir í samskiptum við Rússa" [New Approaches in the Relationship with Russia Need to Be Examined]. *RUV*, March 29, 2022. https://www.ruv.is/frett/2022/03/20/thurfa-ad-skoda-nyjar-adferdir-i-samskiptum-vid-russa.

Printed and Internet Sources

"Address by U.S. Secretary of State Mike Pompeo at the Arctic Council Ministerial Meeting in Rovaniemi, Finland, May 6, 2018." https://www.c-span.org/video/?460478-1/secretary-state-pompeo-warns-russia-china-arctic-policy-address-finland.

"Agreement between the United States of America and Iceland. Effected by the Exchange of Notes at Reykjavik," October 13 and 17, 2017. https://www.state.gov/wp-content/uploads/2019/02/17-1017-Iceland-Defense-Coop-Notes.pdf.

"Agreement on the Platform for the Coalition Government of the Independence Party, the Left Green Movement and the Progressive Party," November 28, 2021. https://www.stjornarradid.is/rikisstjorn/stjornarsattmali/.

"Áhyggjur af norðurljósarannsóknum" [Worries about aurora reseach], *Morgunblaðið*, October 19, 2024. https://www.mbl.is/greinasafn/grein/1873365/?item_num=2&dags=2024-10-19&t=515572980&_t=1729857459.4462316.

Arctic Economic Council. "AEC appoints a new chair from the North Norwegian Company." May 30, 2023. https://arcticeconomiccouncil.com/news/aec-appoints-a-new-chair-from-the-north-norwegian-company/.

Arctic Economic Council. "Banks banning investments in the Arctic will hurt even more now," May 5, 2022. https://arcticeconomiccouncil.com/news/banks-banning-investments-in-the-arctic-will-hurt-even-more-now/.

"Arctic sovereignty a priority: Harper." *CBS*, August 23, 2010. https://www.cbc.ca/news/politics/arctic-sovereignty-a-priority-harper-1.951536

Alfreðsdóttir, Lilja. "Brottför varnarliðsins – þróun varnarmála" [Icelandic National Security in the Post-IDF Era], October 6, 2016. https://www.stjornarradid.is/library/04-Raduneytin/Utanrikisraduneytid/PDF-skjol/Vardberg%2D%2D-raeda-utanri%CC%81kisra%CC%81dherra.pdf.

Arctic Council. *Arctic Human Development Report* (AHDR). Akureyri: Stefansson Arctic Institute, 2004a. file:///C:/Users/vi/Downloads/Arctic%20Human%20Development%20Report.pdf.

Arctic Council. *Impacts of a Warming Arctic: Arctic Climate Change Impact Assessment* (ACIA). Cambridge: Cambridge University Press, 2004b. https://www.amap.no/documents/download/1058/inline.

Arctic Council. "Agreement on Cooperation on Aeronautical and Maritime Search and Rescue in the Arctic," May 12, 2011. https://oaarchive.arctic-council.org/items/9c343a3f-cc4b-4e75-bfd3-4b318137f8a2.

Arctic Council. "Agreement on Marine Oil Pollution Preparedness and Response in the Arctic," May 15, 2013. https://oaarchive.arctic-council.org/items/ee4c9907-7270-41f6-b681-f797fc81659f.

Arctic Council. "New Observer: West Nordic Council, September 11, 2017a, https://arctic-council.org/news/new-observer-west-nordic-council/.
Arctic Council. "Agreement on Enhancing International Arctic Scientific Cooperation," May 11, 2017b. https://oaarchive.arctic-council.org/items/9d1ecc0c-e82a-43b5-9a2f-28225bf183b9
Arctic Council. *Arctic Council Strategic Plan 2021–2030*, May 20, 2021. https://oaarchive.arctic-council.org/handle/11374/2601.
Arctic Council Secretariat. *Kiruna Senior Arctic Officials Report to Minister*, May 15, 2013. https://www.arctic-council.org/index.php/en/document-archive/category/425-main-documents-from-kiruna-ministerial-meeting.
Arctic Council Secretariat. *Annual Report 2013*, April 2014. https://oaarchive.arctic-council.org/handle/11374/941.
Áhættumatsskýrsla fyrir Ísland. Hnattrænir, samfélagslegir og hernaðarlegir þættir [A Risk Assessment Report on Iceland: Global, Societal, and Military Factors]. Reykjavík: Icelandic Ministry for Foreign Affairs, 2009.
Ársskýrsla Orkustofnunar 2013 [The Annual Report of the Icelandic National Energy Authority]. Reykjavík: Orkustofnun 2014. https://rafhladan.is/bitstream/handle/10802/18132/OS-arsskyrsla-2013.pdf?sequence=33
"Ávarp Katrínar Jakobsdóttur forsætisráðherra við opnun Arctic Circle" [Address by PM Katrín Jakobsdóttir at the Opening of the Arctic Circle], October 13, 2022. https://www.stjornarradid.is/efst-a-baugi/frettir/stok-frett/2022/10/13/Forsaetisradherra-flutti-avarp-vid-opnun-things-Hringbords-Nordursloda/.
"Barents Euro-Arctic cooperation: Joint Statement of the European Union, Finland, Denmark, Iceland, Norway and Sweden on suspending activities with Russia," March 9, 2022. https://www.eeas.europa.eu/eeas/barents-euro-arctic-cooperation-joint-statement-european-union-finland-denmark-iceland-norway_en.
Bjarnason, Björn. *Nordic Foreign and Security Policy 2020: Climate Change, Hybrid & Cyber Threats and Challenges to the Multilateral, Rule-Based World Order*, July 2020. https://www.stjornarradid.is/library/02-Rit%2D%2Dskyrslur-og-skrar/NORDIC_FOREIGN_SECURITY_POLICY_2020_FINAL.pdf.
"Comment by Foreign Ministry Spokeswoman Maria Zakharova on the situation around the Northern Dimension and the Barents Euro-Arctic Council (BEAC)," March 11, 2022. https://mid.ru/ru/foreign_policy/news/1803807/?lang=en.
Conference of Parliamentarians of the Arctic Region. https://arcticparl.org/about/.
Conference Proceedings. "Breaking the Ice: Arctic Development and Maritime Transportation: Prospects of the Transarctic Route – Impact and Opportunities."

Akureyri, March 27, 2007. https://www.mfa.is/news-and-publications/nr/3586.

Congressional Research Service. *Changes in the Arctic: Background and Issues for Congress* (updated February 1, 2021). https://crsreports.congress.gov/product/pdf/R/R41153/177.

"Council of the Baltic Sea States Terms of Reference," March 1992. https://cbss.org/wp-content/uploads/2020/06/Council-of-the-Baltic-Sea-States-Terms-of-Reference.pdf.

"Drög að frumvarpi til laga um breytingu á varnarmálalögum nr. 34/2008 (á öryggissvæði o.fl.)" [A Draft of a Legislative Change in the Defense Law], February 26, 2021. https://samradsgatt.island.is/oll-mal/$Cases/Details/?id=2932.

Eide, Espen Barth. "Opening Remarks at Arctic Frontiers: The Arctic – The New Cross-roads," January 21, 2013. https://www.regjeringen.no/en/historical-archive/Stoltenbergs-2nd-Government/Ministry-of-Foreign-Affairs/taler-og-artikler/2013/remarks_frontiers/id713462.

Embassy of China in Iceland. "Sino-Icelandic Economic and Trade Relationship." Reykjavík: Chinese Embassy, 2019. https://is.china-embassy.gov.cn/eng/zbgx/jmgx/201904/t20190410_3164334.htm.

Embassy of Iceland in Beijing. Press Release, "China and Iceland sign agreements on geothermal and geoscience cooperation and in the field of polar affairs," April 23, 2012. https://www.iceland.is/iceland-abroad/cn/english/news-and-events/china-and-iceland-sign-agreements-on-geothermal-and-geoscience-cooperation-and-in-the-field-of-polar-affairs/8882/.

European Parliament Resolution on Arctic governance, October 9, 2008. https://www.europarl.europa.eu/doceo/document/TA-6-2008-0474_EN.html.

Expert Commission on Norwegian Security and Defense Policy. *Unified Effort*. Oslo: Norwegian Ministry of Defense, 2015. https://www.regjeringen.no/globalassets/departementene/fd/dokumenter/unified-effort.pdf.

"First Joint Meeting between the Arctic Council and the Arctic Economic Council." Arctic Council, October 19, 2019. https://arctic-council.org/news/first-joint-meeting-between-the-arctic-council-and-the-arctic-economic-council/#:~:text=The%20Arctic%20Council%20and%20the%20Arctic%20Economic%20Council%20hold%20their,Participants%2C%20and%20the%20Councils'%20respective.

Garðarsson, Guðmundur. Þingræða [Parliamentary speech], April 25, 1989. *Þingtíðindi* [Icelandic parliamentary records]. https://www.althingi.is/altext/111/r3/3675.html.

Gísladóttir, Ingibjörg Sólrún. "Opening Address at the Symposium of the Law of the Sea Institute of Iceland on the Legal Status of the Arctic Ocean, delivered by the Minister for Foreign Affairs." Reykjavík, November 9, 2007. https://www.mfa.is/news-and-publications/nr/3983.

Gunnarsson, Einar. "Together Towards a Sustainable Arctic: One Year into the 2019–2021 Icelandic Chairmanship." Arctic Council, June 9, 2020. https://www.arctic-council.org/news/one-year-into-the-2019-2021-icelandic-chairmanship/.

Gunnarsson, Einar. "Iceland has worked to address priority challenges–economic growth, social inclusion and environmental protection–during its chairmanship of the Arctic Council." *The Foreign Service Journal*, May 2021. https://afsa.org/toward-sustainable-arctic.

Gorbachev, Mikhail. "The Speech in Murmansk at the ceremonial meeting on the occasion of the presentation of the Order of Lenin and the Gold Star Medal in the city of Murmansk, October 1, 1987." https://www.barentsinfo.fi/docs/Gorbachev_speech.pdf.

Hagstofa Íslands [Statistics Iceland] . "Vöruviðskipti óhagstæð um 363,3 milljarða árið 2023" [The Trade Imbalance Amounts to 363.3 billion Icelandic Kronur in 2023]. https://hagstofa.is/utgafur/frettasafn/utanrikisverslun/voruvidskipti-i-arid-2023-lokatolur/.

Hannibalsson, Jón Baldvin. Þingræða fjármálaráðherra [A parliamentary speech by the Icelandic Finance Minister]. *Þingtíðindi*, February 26, 1988. https://www.althingi.is/altext/raeda/?rnr=3418<hing=110.

Hermannsson, Steingrímur. "Ræða Steingríms Hermannssonar utanríkisráðherra um utanríkismál" [Speech by Foreign Minister Steingrímur Hermannsson on Foreign Affairs]. *Þingtíðindi*, February 22, 1987. https://www.althingi.is/altext/110/s/pdf/0596.pdf.

Icelandic Ministry for Foreign Affairs. *Iceland's Participation in International Crisis Management and Peacekeeping*," November 2000. Reykjavík: Icelandic Ministry for Foreign Affairs.

Icelandic Ministry for Foreign Affairs. *North Meets North: Navigation and the Future of the Arctic*, July 2006 [Translated from the original report in Icelandic entitled *Fyrir stafni haf. Tækifæri tengd siglingum á norðurslóðum* [Ocean Ahead: Opportunities Linked to Arctic Shipping], February 2005]. https://www.stjornarradid.is/media/utanrikisraduneyti-media/media/Utgafa/vef_skyrsla.pdf.

Icelandic Ministry for Foreign Affairs, *Ísinn brotinn. Þróun norðurskautssvæðisins og sjóflutningar, horfur í siglingum á Norður-Íshafsleiðinni*, October 2006a [Broken Ice: Arctic Developments and Sea Transports; Prospects for Arctic Shipping]. https://www.stjornarradid.is/media/utanrikisraduneyti-media/media/utgafa/isinn_brotinn.pdf.

Icelandic Ministry for Foreign Affairs. "Samkomulag Bandaríkjanna og Íslands um aðgerðir til að styrkja varnarsamstarf ríkjanna" [Joint Understanding between the United States and Iceland on Measures to Strengthen the Bilateral Defense Cooperation], October 2006b. https://www.utanrikisraduneyti.is/media/Frettatilkynning/Samkomulag_um_varnarmal.pdf.

Icelandic Ministry for Foreign Affairs. "Samkomulag um samstarf á sviði öryggismála, varnarmála og viðbúnaðar milli Noregs og Íslands" [Memorandum of Understanding between Norway and Iceland on Security, Defense, and Preparedness], April 2007a. https://www.utanrikisraduneyti.is/media/Frettatilkynning/MOU_-_undirritun.pdf.

Icelandic Ministry for Foreign Affairs. "Yfirlýsing lýðveldisins Íslands og konungsríkisins Danmerkur um samstarf í víðari skilningi um öryggis- og varnarmál og almannavarnir" [Joint Declaration between the Republic of Iceland and the Kingdom of Denmark on Broader Cooperation on Security and Defense and Civil Preparedness], April 2007b. https://www.utanrikisraduneyti.is/media/Frettatilkynning/Yfirlysing_Islands_og_Danmerkur.pdf.

Icelandic Ministry for Foreign Affairs. "Samkomulag um samstarf á sviði varnar- og öryggismála milli breska konungsríkisins og Íslands" [Agreement between the United Kingdom and Iceland on Defense and Security Cooperation], May 2008. https://www.utanrikisraduneyti.is/media/PDF/UK-Iceland_MoU_-Icelandic.pdf.

Icelandic Ministry for Foreign Affairs. Press Release, "NATO summit welcomes Iceland's initiative on the High North," April 6, 2009a. https://www.mfa.is/news-and-publications/nr/4926.

Icelandic Ministry for Foreign Affairs. "Samkomulag milli utanríkisráðuneytis Íslands og varnarmálaráðuneytis Kanada um samstarf í varnarmálum" [Agreement between the Icelandic Foreign Ministry and the Canadian Defense Ministry on Defense Cooperation], October 14, 2010. https://www.stjornarradid.is/media/utanrikisraduneyti-media/media/Frettatilkynning/Ice-Can-MOU-final-Icelandic.PDF.

Icelandic Ministry for Foreign Affairs. Report, *Ísland á norðurslóðum* [Iceland in the Arctic], April 2009b. https://www.utanrikisraduneyti.is/media/Skyrslur/Skyrslan_Island_a_nordurslodumm.pdf.

Icelandic Ministry for Foreign Affairs. Press Release, "Launch of Icelandic Arctic Cooperation Network and arrival of first Nansen Professor," February 12, 2013a. https://www.government.is/news/article/2013/02/12/Launch-of-Icelandic-Arctic-Cooperation-Network-and-arrival-of-first-Nansen-Professor/.

Icelandic Ministry for Foreign Affairs. "The Arctic as Global Challenge – Issues and Solution." Speech by Foreign Minister Össur Skarphéðinsson, March 18, 2013b. https://www.mfa.is/media/Raedur/The-Arctic-as-a-Global-Challenge%2D%2D-Speech-by-Ossur-Skarphedinsson.pdf.

Icelandic Ministry for Foreign Affairs. "Free Trade Agreement between Iceland and China," April 15, 2013c. https://www.mfa.is/foreign-policy/trade/free-trade-agreement-between-iceland-and-china/.

Icelandic Ministry for Foreign Affairs. "Free Trade Agreement between Iceland and China." Fact Sheet, April 15, 2013d. https://www.mfa.is/media/fta kina/China_fact_sheet_enska_15042013.

Icelandic Ministry for Foreign Affairs. "Vegna '5-ríkja samráðs' um fiskveiðar í Norður-Íshafi" [On "Five-State Consultation" on Fishing in the Arctic Ocean], July 23, 2015. https://www.stjornarradid.is/efst-a-baugi/frettir/stok-frett/2015/07/23/Vegna-5-rikja-samrads-um-fiskveidar-i-Nordur-Ishafi-nbsp/.

Icelandic Ministry for Foreign Affairs. Report, *Áfram gakk. Utanríkisviðskiptastefna Íslands* [March Forward: Iceland's External Trade Policy], December 2020a. https://www.stjornarradid.is/library/04-Raduneytin/Utanrikisraduneytid/PDF-skjol/Sky%cc%81rsla%20-%20A%cc%81fram%20gakk!%20Utanri%cc%81kisvi%c3%b0skiptastefna%20I%cc%81slands%20-%20Copy%20(1).pdf.

Icelandic Ministry for Foreign Affairs. "Auglýsing um samning um að koma í veg fyrir stjórnlausar úthafsveiðar í miðhluta Norður-Íshafsins" [Announcement on the Agreement on the Prevention of Unregulated High Seas Fishing in the Central Arctic Ocean], June 22, 2021a. https://www.stjornarradid.is/library/03-Verkefni/Utanrikismal/Lagamal/C-deildar-atak/16_2021_Augl%C3%BDsing%20um%20samning%20um%20a%C3%B0%20koma%20%C3%AD%20veg%20fyrir%20stj%C3%B3rnlausar%20%C3%BAthafsvei%C3%B0ar%20%C3%AD%20mi%C3%B0hluta%20Nor%C3%B0ur-%C3%8Dshafsins..pdf.

Icelandic Ministry for Foreign Affairs. *Samstarf Grænlands og Íslands á nýjum Norðurslóðum. Tillögur Grænlandsnefndar utanríkis- og þróunarsamvinnuráðherra* [Cooperation between Greenland and Iceland in the New Arctic. Proposals by the Greenland Committee of the Minister for Foreign Affairs and International Development Cooperation], December 2020b. https://www.stjornarradid.is/library/02-Rit%2D%2Dskyrslur-og-skrar/Samstarf%20Gr%c3%a6nlands%20og%20%c3%8dslands%20%c3%a1%20n%c3%bdjum%20Nor%c3%b0ursl%c3%b3%c3%b0um%20-%20Copy%20(1).pdf

Icelandic Ministry for Foreign Affairs. *Norðurljós. Skýrsla starfshóps um efnahagstækifæri á norðurslóðum* [Northern Lights, a Working Group Report about Economic Opportunities in the Arctic], May 2021b. https://www.stjornarradid.is/library/04Raduneytin/Utanrikisraduneytid/PDF-skjol/Nor%C3%B0urlj%C3%B3s%20-%20efnahagst%C3%A6kif%C3%A6ri%20%C3%A1%20nor%C3%B0ursl%C3%B3%C3%B0um_WEB.pdf.

Icelandic Ministry for Foreign Affairs. "Service visits of US submarines authorized," April 18, 2023a. https://www.government.is/diplomatic-missions/embassy-article/2023/04/18/Service-visits-of-US-submarines-authorised/#:~:text=The%20Minister%20for%20Foreign%20Affairs,supplies%20and%20exchange%20crew%20members.

Icelandic Ministry for Foreign Affairs. "Iceland suspends embassy operations in Moscow," June 9, 2023b. https://www.government.is/diplomatic-missions/embassy-article/2023/06/09/Iceland-suspends-embassy-operations-in-Moscow/.

Icelandic Prime Minister's Office. *The Economic Impact of the Russian Counter-Sanctions on Trade between Iceland and the Russian Federation*, January 2016. https://www.stjornarradid.is/media/forsaetisraduneyti-media/media/skyrslur/theeconomicimpactoftherussiansanctionsontradebetweenicelandandrussia.pdf.

Icelandic Prime Minister's Office. *Hagsmunir Íslands á norðurslóðum. Tækifæri og áskoranir* [Iceland's Interests in the Arctic: Opportunities and Challenges], 2020. https://www.stjornarradid.is/media/forsaetisraduneyti-media/media/skyrslur/hagsmunamat_skyrsla-lr-.pdf.

Icelandic Parliament. Alþingistíðindi [Parliamentary Records], 1954, D umræður um þingsályktunartillögur og fyrirspurnir [Parliamentary Debates and Questions], Reykjavík: Gutenberg Printing House, 1956.

Icelandic Parliament. *Þingskjöl* [Parliamentary Documents] 1954 A. Reykjavík: Gutenberg Printing House, 1955.

Icelandic Parliament. "Svar utanríkisráðherra við fyrirspurn frá Andrési Inga Jónssyni um viðveru herliðs á Keflavíkurflugvelli [The Foreign Minister's [Guðlaugur Þór Þórðarson] Response to a Question by Andrés Ingi Jónsson, on [Foreign] Military Presence at Keflavík Airport]," November 14, 2018. https://www.althingi.is/altext/149/s/0427.htm.

Icelandic Parliament. "Þingsályktun um fullgildingu samnings um að koma í veg fyrir stjórnlausar úthafsveiðar í miðhluta Norður-Íshafsins" [Approval of a Parliamentary Proposal on the Ratification of the Agreement on the Prevention of Irregular High Seas Fishing in the Central Arctic Ocean], June 3, 2019. https://www.althingi.is/altext/149/s/1685.html.

Icelandic Prime Minister's Office, "Declaration of Cooperation between the Prime Minister of Greenland and the Prime Minister of Iceland on Future Cooperation between Greenland and Iceland," October 13, 2022. https://www.government.is/library/01-Ministries/Prime-Ministrers-Office/Declaration%20of%20cooperation.pdf.

Icelandic Prime Minister's Office and the Ministry for Foreign Affairs. "Ísland fordæmir innrás Rússa í Úkraínu" [Iceland Condemns the Invasion of Russia into Ukraine], February 24, 2022. https://www.stjornarradid.is/efst-a-baugi/frettir/stok-frett/2022/02/24/Island-fordaemir-innras-Russa-i-Ukrainu.

"Innrás Rússlands í Úkraínu – viðbrögð íslenskra stjórnvalda: íslensk stjórnvöld fordæma ólögmæta innrás Rússlands og lýsa yfir algjörum stuðningi við Úkraínu" [Russia's Invasion of Ukraine—Iceland's Reaction: The Icelandic Government Condemns Russia's Illegal Invasion and Declares Full Support for Ukraine]. The Icelandic Government, October 6, 2022. https://www.stjornarradid.is/verkefni/utanrikismal/strid-i-ukrainu-vidbrogd-islenskra-stjornvalda/.

"Joint Declaration between the Department of Defense of the United States of America and the Ministry for Foreign Affairs of Iceland," June 29, 2016.

https://www.stjornarradid.is/library/04-Raduneytin/Utanrikisraduneytid/PDF-skjol/Joint-Declaration%2D%2DSigned-.PDF.

Jónsson, Benedikt. "Address by Ambassador Benedikt Jónsson on behalf of Iceland's Chairmanship of the Arctic Council," at the International Round Table – Indigenous Peoples of the North and the Parliamentary System of the Russian Federation: Experience and Prospects, Moscow, March 12–13, 2003. https://www.utanrikisraduneyti.is/frettaefni/ymis-erindi/nr/224.

Letter, John Moloolenar and Raja Krishnamoorthi to Antony Blinken and Lloyd Austin, October 16, 2024. https://selectcommitteeontheccp.house.gov/sites/evo-subsites/selectcommitteeontheccp.house.gov/files/evo-media-document/10.16.24_PRC%20dual%20use%20research%20in%20the%20Arctic.

"A mad scramble for the shrinking Arctic." *New York Times*, September 10, 2008. https://www.nytimes.com/2008/09/10/opinion/10iht-edarctic.1.16040367.html.

Ministry of Foreign Affairs of the Russian Federation. "Foreign Ministry Statement on the withdrawal of the Russian Federation from the Council of the Baltic Sea States," May 17, 2022. https://www.mid.ru/en/foreign_policy/news/1813674/.

Ministry of Foreign Affairs of the Russian Federation. "Foreign Minister Sergey Lavrov's video address to the participants in the 13th session of the Arctic Council, May 11, 2023a." https://mid.ru/en/foreign_policy/rso/1869388/.

Ministry of Foreign Affairs of the Russian Federation. "Foreign Ministry Statement on Russia's withdrawal from the Barents Euro-Arctic Council," September 18, 2023b. https://mid.ru/en/foreign_policy/rso/1904899/.

NATO. "Pre-ministerial press conference by NATO Secretary General Jens Stoltenberg ahead of the meetings of NATO Defence Ministers." October 11, 2023. https://www.nato.int/cps/en/natohq/opinions_208037.htm.

"NATO 2022 Strategic Concept," March 3, 2023. https://www.nato.int/cps/en/natohq/topics_210907.htm#:~:text=The%202022%20Strategic%20Concept%20describes,and%20management%3B%20and%20cooperative%20security.

Nordic Council. "Nordic Council calls for closer co-operation on foreign and security policy," February 9, 2021. https://www.norden.org/en/news/nordic-council-calls-closer-co-operation-foreign-and-security-policy.

Norwegian Ministry of Foreign Affairs. *Regjeringens nordområdestrategi* [The Government's High North Strategy]. Oslo: Norwegian Ministry of Foreign Affairs, 2006.

Norwegian Ministry of Foreign Affairs. *Mennesker, muligheter og norske interesser i nord* [People, Opportunities and Norwegian Interests in the North], November 27, 2020. https://www.regjeringen.no/no/dokumenter/meld.-st.-9-20202021/id2787429/.

"Northern Dimension Policy: Joint Statement by the European Union, Iceland and Norway on suspending activities with Russia and Belarus," March 8, 2022. https://ndpculture.org/news/northern-dimension-policy-joint-statement-by-the-european-union-iceland-and-norway-on-suspending-activities-with-russia-and-belarus/.

Organization for Economic Co-operation and Development (OECD). "OECD Benchmark Definition of Foreign Direct Investment – Fourth Edition," 2008. https://www.oecd.org/daf/inv/investmentstatisticsandanalysis/40193734.pdf.

"Outlines of the Statement by Sergey Lavrov at the 12th Ministerial Meeting of the Arctic Council." Reykjavík, May 20, 2021. https://oaarchive.arctic-council.org/bitstream/handle/11374/2683/Statement%20by%20Minister%20S.Lavrov.pdf?sequence=1&isAllowed=; https://apnews.com/article/donald-trump-iceland-europe-russia-middle-east-24e823ef46207bfbe4c841e1b4a864f0y.

Parliamentary speech by Halldór Ásgrímsson. "Orð forseta um Evrópusambandið" [The President's Comments on the European Union]. *Þingtíðindi*, April 17, 2002. https://www.althingi.is/altext/127/04/r17104201.sgml.

Pálsson, Gunnar. "The Icelandic Chairmanship Program," Address by Ambassador Gunnar Pálsson, Chair of Senior Arctic Officials, Northern Forum 6th General Assembly, St. Petersburg, April 24, 2003a. https://www.stjornarradid.is/efst-a-baugi/frettir/stok-frett/2003/04/25/Aaetlun-Islands-i-formennsku-Nordurskautsradsins-2002-2004/.

Pálsson, Gunnar. "Arctic co-operation 12 years on: How successful?" Address by the Chairman of Senior Arctic Officials Wilton Park Conference, United Kingdom, March 20, 2003b. https://www.utanrikisraduneyti.is/frettaefni/ymis-erindi/nr/235.

Permanent Mission of Iceland to the United Nations. "Iceland and the United Nations," n.d., accessed February 10, 2024, https://www.government.is/diplomatic-missions/permanent-mission-of-iceland-to-the-united-nations/iceland-and-the-united-nations/.

Pompeo, Mike. "Looking North: Sharpening America's Arctic Focus." U.S. Department of State, May 6, 2019. https://2017-2021.state.gov/looking-north-sharpening-americas-arctic-focus/.

Press Release. "Judgment in Case E-16/11 EFTA Surveillance Authority v Iceland ('Icesave'). Application of the EFTA Surveillance Authority in the Case of Icesave Dismissed." EFTA Court, January 28, 2013. https://eftacourt.int/wp-content/uploads/2019/01/16_11_PR_EN1.pdf?x85212.

"Ræða Davíðs Oddssonar utanríkisráðherra um utanríkismál" [Address by Foreign Minister Davíð Oddsson on Foreign Affairs to the Icelandic Parliament], November 11, 2004. https://www.stjornarradid.is/raduneyti/utanrikisraduneytid/utanrikisradherra/fyrri-radherrar/stok-raeda-fyrrum-radherra/2004/11/11/Raeda-Davids-Oddssonar-utanrikisradherra-um-utanrikismal/.

"Ræða Davíðs Oddssonar utanríkisráðherra um utanríkismál" [Address by Foreign Minister Davíð Oddsson on Foreign Affairs to the Icelandic Parliament], April 29, 2005. https://www.stjornarradid.is/efst-a-baugi/frettir/stok-frett/2005/04/29/Raeda-Davids-Oddssonar-utanrikisradherra-um-utanrikismal/.

"Ræða Össurar Skarphéðinssonar utanríkisráðherra um utanríkismál" á Alþingi [Parliamentary Address by Foreign Minister Össur Skarphéðinsson], *Þingtíðindi*, May 14, 2010, accessed May 20, 2023, https://www.althingi.is/altext/raeda/138/rad20100514T105208.html.

"Ræða utanríkisráðherra [Þórdís Kolbrún Reykfjörð Gylfadóttir] á Helsinki Security Forum á þingi Hringborðs norðurslóða" [Address by the Icelandic Foreign Minister at the Arctic Circle's Helsinki Security Forum]," October 17, 2022. https://www.stjornarradid.is/raduneyti/utanrikisraduneytid/utanrikisradherra/stok-raeda-utanrikisradherra/2022/10/17/Raeda-utanrikisradherra-a-Hringbordi-nordursloda-Arctic-Circle/.

Reagan, Ronald. "President Reagan's written responses to questions submitted by the Finnish newspaper *Helsingin Sanomat*," May 19, 1988. The American Presidency Project, UC Santa Barbara. https://www.presidency.ucsb.edu/documents/written-responses-questions-submitted-the-finnish-newspaper-helsingin-sanomat.

"Reykjavík Declaration." Arctic Council, May 20, 2021. https://oaarchive.arctic-council.org/handle/11374/2600.

"Samstarfsyfirlýsing ríkisstjórnar 2009" [The Platform of the Coalition Government [between the Social Democratic Alliance and the Left Greens] 2009], May 10, 2009. https://www.stjornarradid.is/rikisstjorn/sogulegt-efni/um-rikisstjorn/2009/05/10/Samstarfsyfirlysing-rikisstjornar-2009/.

"A scramble for the Arctic: With one fifth of the world's oil and gas at stake, countries are struggling to control the once-frozen arctic." *Aljazeera*, August 12, 2010. https://www.aljazeera.com/features/2010/12/8/a-scramble-for-the-arctic;

"Scramble for the Arctic." *Financial Times*, August 19, 2007. https://www.ft.com/content/65b9692c-4e6f-11dc-85e7-0000779fd2ac.

Sigurðardóttir, Jóhanna. "Beðist afsökunar á vanrækslu og andvaraleysi" [An Apology for Failures and Complacency]. Prime Minister's Office, October 6, 2009. https://www.stjornarradid.is/efst-a-baugi/frettir/stok-frett/2009/10/06/Bedist-afsokunar-a-vanraekslu-og-andvaraleysi/.

Skarphéðinsson, Össur. "The Arctic as Global Challenge – Issues and Solution." Speech by the Minister for Foreign Affairs, March 18, 2013b. https://www.mfa.is/media/Raedur/The-Arctic-as-a-Global-Challenge%2D%2D-Speech-by-Ossur-Skarphedinsson.pdf.

Skarphéðinsson, Össur. "Framsöguræða Össurar Skarphéðinssonar utanríkisráðherra með skýslu til Alþingis um utanríkis- og alþjóðamál" [Address of the Minister for Foreign Affairs, when presenting his Annual Report to the Parliament], February 14, 2013c.

Skarphéðinsson, Össur. Þingræða [Parliamentary speech], September 3, 2010, accessed February 9, 2024. *Þingtíðindi*. https://www.althingi.is/altext/raeda/138/rad20100903T111357.html.

Skýrsla Halldórs Ásgrímssonar utanríkisráðherra um utanríkis-og alþjóðamál [Report by the Minister for Foreign Affairs on Foreign and International Affairs], *Þingtíðindi*, April 2004. https://www.althingi.is/altext/130/s/pdf/1377.pdf.

Skýrsla Ingibjargar Sólrúnar Gísladóttur utanríkisráðherra um utanríkis- og alþjóðamál [Report by Foreign Minister Ingibjörg Sólrún Gísladóttir on Foreign and International Affairs]. *Þingtíðindi*, April 2008. https://www.althingi.is/altext/135/s/0857.html.

Skýrsla Össurar Skarphéðinssonar utanríkisráðherra um utanríkis- og alþjóðamál, *Þingtíðindi*, maí 2010. https://www.althingi.is/altext/138/s/1070.html

Skýrsla Össurar Skarphéðinssonar utanríkisráðherra um utanríkis- og alþjóðamál [Report by Foreign Minister Össur Skarphéðinsson on Foreign and International Affairs]. *Þingtíðindi*, April 2012. https://www.althingi.is/al*text/140/s/pdf/1229.pdf.

Skýrsla Gunnars Braga Sveinssonar utanríkisráðherra um utanríkis- og alþjóðamál [Report by Foreign Minister Gunnar Bragi Sveinsson on Foreign and International Affairs]. *Þingtíðindi*, March 2014. https://www.althingi.is/altext/143/s/0757.html.

Skýrsla Gunnars Braga Sveinssonar utanríkisráðherra um utanríkis- og alþjóðamál [Report by Foreign Minister Gunnar Bragi Sveinsson on Foreign and International Affairs]. *Þingtíðindi*, March 2015. https://www.althingi.is/altext/144/s/1074.html.

Skýrsla Gunnars Braga Sveinssonar utanríkisráðherra um utanríkis- og alþjóðamál [Report by Foreign Minister Gunnar Bragi Sveinsson on Foreign and International Affairs]. *Þingtíðindi*, March 2016. https://www.althingi.is/altext/145/s/1009.html.

Skýrsla Guðlaugs Þórs Þórðarsonar utanríkisráðherra um utanríkis- og alþjóðamál [Report by Foreign Minister Guðlaugur Þór Þórðarsonar on Foreign and International Affairs]. *Þingtíðindi*, May 2017. https://www.althingi.is/altext/146/s/0671.html.

Skýrsla Guðlaugs Þórs Þórðarsonar utanríkisráðherra um utanríkis- og alþjóðamál [Report by Foreign Minister Guðlaugur Þór Þórðarson on Foreign and International Affairs], April 2019. https://www.stjornarradid.is/gogn/rit-og-skyrslur/stakt-rit/2019/04/29/Skyrsla-utanrikisradherra-um-utanrikis-og-althjodamal-logd-fyrir-Althingi/.

Skýrsla Guðlaugs Þórs Þórðarsonar utanríkis- og þróunarsamvinnuráðherra um utanríkis- og alþjóðamál [Report by Minister for Foreign Affairs and Development Cooperation, Guðlaugur Þór Þórðarson on Foreign and International Affairs], May 2020. https://www.stjornarradid.is/gogn/rit-og-skyrslur/stakt-rit/2020/05/07/Skyrsla-utanrikis-og-throunarsamvinnuradherra-um-utanrikis-og-althjodamal-2020/.

Skýrsla Guðlaus Þórs Þórðarsonar utanríkis- og þróunarsamvinnuráðherra um utanríkis- og alþjóðamál [Report by Minister for Foreign Affairs and Development Cooperation, Guðlaugur Þór Þórðarson on Foreign and International Affairs], May 2021. https://www.stjornarradid.is/efst-a-baugi/frettir/stok-frett/2021/05/05/Skyrsla-utanrikis-og-throunarsamvinnuradherra-um-utanrikis-og-althjodamal-2021/.

Skýrsla Þórdísar Kolbrúnar Reykfjörð Gylfadóttur um utanríkis- og alþjóðamál [Report by Minister for Foreign Affairs, Þórdís Kolbrún Reykfjörð Gylfadóttir on Foreign and International Affairs], May 2024. https://www.stjornarradid.is/gogn/rit-og-skyrslur/stakt-rit/2024/05/14/Skyrsla-utanrikisradherra-um-utanrikis-og-althjodamal/.

"Statement of the Government of the People's Republic of China and the Government of the Kingdom of Norway on Normalization of Bilateral Relations." December 19, 2016. https://www.regjeringen.no/globalassets/departementene/ud/vedlegg/statement_kina.pdf.

"Statement published by the Norwegian MFA: Russia Suspended from the Council of the Baltic Sea States," March 3, 2022. https://www.regjeringen.no/en/aktuelt/russland-suspenderes-fra-ostersjoradet/id2903009/.

"Stefnuyfirlýsing ríkisstjórnar [Sjálfstæðisflokks og Samfylkingar] 2007" [The Government Platform of the Independence Party and the Social Democratic Alliance, 2007], May 22, 2007. https://www.stjornarradid.is/media/stjornarrad media/media/Rikjandi_rikisstjorn/stefnuyfirlysing-23-3-2013.pdf.

"Stefnuyfirlýsing ríkisstjórnar [Framsóknarflokks og Sjálfstæðisflokks] [The Government Platform of the Progressive Party and the Independence Party], May 23, 2013. https://www.government.is/government/coalition-platform/http:/.

Stoltenberg, Thorvald. *Nordic cooperation on foreign and security policy. Proposals presented to the extraordinary meeting of Nordic foreign ministers in Oslo*, February 9, 2009. https://www.regjeringen.no/globalassets/upload/ud/vedlegg/nordicreport.pdf.

Støre, Jonas Gahr. "Iceland and Norway – Neighbours in the High North." Speech delivered at the University of Iceland, November 3, 2008. https://www.regjeringen.no/en/dep/ud/Whats-new/Speeches-and-articles/speeches_foreign/2008/iceland-and-norway%2D%2Dneighbours-in-the-hi.html?id=534706.

Sveinsson, Gunnar Bragi. "Iceland's Role in the Arctic – The Future of Arctic Cooperation." Arctic Circle Assembly Reykjavík, October 14, 2013. https://www.stjornarradid.is/media/utanrikisraduneyti-media/media/nordurslodir/Arctic-Circle-speech-October-14-2013.pdf.

Sveinsson, Gunnar Bragi. "Iceland's Policy and Priorities in a Changing Arctic," January 20, 2014. https://www.stjornarradid.is/media/utanrikisraduneyti-media/media/Raedur/Raeda-GBS-ArcticFrontiers.pdf.

Sverrisdóttir, Valgerður. Address by the Minister for Foreign Affairs at the Conference "Breaking the Ice: Arctic Development and Maritime Transportation: Prospects of the Transarctic Route – Impact and Opportunities." Akureyri, March 27, 2007. https://www.mfa.is/news-and-publications/nr/3586.

Tillögur nefndar um endurskoðun á stefnu Íslands í málefnum norðurslóða [Proposals of a [Parliamentary] Committee on the Revision of Iceland's Arctic Policy], March 8, 2021. https://www.stjornarradid.is/library/04-Raduneytin/Utanrikisraduneytid/PDF-skjol/Skilabréf%20og%20tillögur%20nefndar%20um%20endurskoðun%20norðurslóðastefnu.pdf.

State Council of the People's Republic of China. *China's Arctic Policy*, January 26, 2018. https://english.www.gov.cn/archive/white_paper/2018/01/26/content_281476026660336.htm.

United Nations, Division for Ocean Affairs and Law of the Sea. "Commission on the Limits of the Continental Shelf (CLCS]. Outer limits of the continental shelf beyond 200 nautical miles from the baselines: Submissions to the Commission: Partial revised Submission by Iceland," April 4, 2021. https://www.un.org/depts/los/clcs_new/submissions_files/submission_isl_rev2021.htm.

U.S. Air Forces in Europe and Air Forces Africa Public Affairs. "U.S. Air Force B-2 Spirit stealth bomber aircraft arrive in Iceland for ally, partner training," August 24, 2021. https://www.usafe.af.mil/News/Press-Releases/Article/2743096/us-air-force-b-2-spirit-stealth-bomber-aircraft-arrive-in-iceland-for-ally-part/.

U.S. Department of Defense. *Arctic Strategy*, November 2013. https://www.hsdl.org/?abstract&did=747036.

U.S. Department of Defense Arctic Strategy, June 2019. https://dod.defense.gov/Portals/1/Documents/pubs/Report_to_Congress_on_Resourcing_the_Arctic_Strategy.pdf.

U.S. Government. *Directive on Arctic Region Policy*, January 9, 2009. PPP-2008-book2-doc-pg1545 (3).pdf

United States Coast Guard. *Arctic Strategic Outlook*. Washington, D.C., 2019. https://www.uscg.mil/Portals/0/Images/arctic/Arctic_Strategic_Outlook_APR_2019.pdf.

U.S. Department of State. *Limits in the Seas, No. 112, United States Responses to Excessive National Maritime Claims*" (Bureau of Oceans and International Environmental and Scientific Affairs, 1992), 1–88.

U.S. Department of Defense. *Summary of the 2018 National Defense Strategy: Sharpening the American Military Competitive Edge*. Washington D.C.: Department of Defense, 2018. https://dod.defense.gov/Portals/1/Documents/pubs/2018-National-Defense-Strategy-Summary.pdf.

U.S. Department of State. "Remarks with Icelandic Minister of Foreign Affairs Ingibjörg Solun Gisladottir," May 30, 2008. https://20012009.state.gov/secretary/rm/2008/05/105447.htm.

U.S. Department of State. "Joint Statement on Arctic Council Following Russia's Invasion of Ukraine" [The United States, Canada, Norway, Denmark, Finland, Sweden, and Iceland], March 3, 2022a. https://www.state.gov/joint-statement-on-arctic-council-cooperation-following-russias-invasion-of-ukraine/.

U.S. Department of State. "Joint Statement on Limited Resumption of Arctic Council Cooperation [The United States, Canada, Norway, Denmark, Finland, Sweden, and Iceland].

U.S. Department of State, June 8, 2022b. https://www.state.gov/joint-statement-on-limited-resumption-of-arctic-council-cooperation/.

U.S. Embassy in Copenhagen. "Statement of Intent on Defense Investments in Greenland," September 17, 2018. https://twitter.com/usembdenmark/status/1041695240686632960?lang=en.

U.S. Geological Survey, *Circum-Arctic Resource Appraisal: Estimates of Undiscovered Oil and Gas North of the Arctic Circle*. Washington, D.C.: U.S. Government, 2008. https://pubs.usgs.gov/fs/2008/3049/.

Wammen, Nikolai, Carl Hagland, Gunnar Bragi Sveinsson, Ine Eriksen Søreide, and Peter Hultqvist. "Við aukum norrænt samstarf á sviði varnarmála" [We Will Increase Nordic Defense Cooperation], April 10, 2015. https://www.stjornarradid.is/efst-a-baugi/frettir/stok-frett/2015/04/10/Vid-aukum-norraent-samstarf-a-svidi-varnarmala/.

Þingsályktun um stefnu Íslands í málefnum norðurslóða [Parliamentary Resolution on Iceland's Policy on the Arctic]. *Þingtíðindi*, May 19, 2011. https://www.althingi.is/altext/139/s/1148.html.

"Þingsályktunartillaga um aukið samstarf Íslands og Grænlands" [Parliamentary Resolution on Increased Cooperation between Iceland and Greenland], May 31, 2021. https://www.althingi.is/altext/151/s/1560.html.

Þjóðaröryggisstefna Íslands [Icelandic National Security Policy], Alþingi [Icelandic Parliament]. *Þingtíðindi*, April 14, 2016. https://www.althingi.is/thingstorf/thingmalalistar-eftir-thingum/ferill/?ltg=145&mnr=327.

Þórðarson, Guðlaugur Þór. "Arctic: Territory of Dialogue." Address at the International Arctic Forum in Arkhangelsk, March 29, 2017. https://www.

stjornarradid.is/raduneyti/utanrikisraduneytid/utanrikisradherra/stok-raeda-utanrikisradherra/2017/03/29/Avarp-a-International-Arctic-Forum-i-Arkhangelsk/.

Þórðarson, Guðlaugur Þór. "Back to the Future: The Geopolitical Centrality of the North Atlantic and the Arctic." Address delivered at the Center for Strategic Studies, Washington, D.C.), May 16, 2018a. https://www.stjornarradid.is/raduneyti/utanrikisraduneytid/utanrikisradherra/stok-raeda-utanrikisradherra/2018/05/16/Raeda-radherra-hja-hugveitunni-CSIS-i-Washington-DC/.

Þórðarson, Guðlaugur Þór. "Iceland-China relations will continue to strengthen." *China Daily*, September 6, 2018b. https://www.chinadaily.com.cn/a/201809/06/WS5b90702ba31033b4f465477b.html.

Þórðarson, Guðlaugur Þór. "New Geo-Political Reality in the West Nordic Area." A conference speech at the Arctic Circle, Reykjavík, October 20, 2018c. https://www.stjornarradid.is/raduneyti/utanrikisraduneytid/utanrikisradherra/stok-raeda-utanrikisradherra/2018/10/20/Raeda-radherra-i-vidburdi-Vestnorraena-radsins-a-Hringbordi-nordursloda-um-sviptingar-i-althjodastjornmalum-og-ahrif-theirra-a-vestnorraena-svaedid/.

Þórðarson, Guðlaugur Þór. "Presentation of the [Arctic Council] Icelandic Chairmanship Program," Arctic Council Ministerial Meeting, Rovaniemi, May 7, 2019a. https://oaarchive.arctic-council.org/handle/11374/2397.

Þórðarson, Guðlaugur Þór. "Statement by the Minister of Foreign Affairs of Iceland," Arctic Council Ministerial Meeting, Rovaniemi, May 7, 2019b. https://oaarchive.arctic-council.org/handle/11374/2402?show=full.

Þórðarson, Guðlaugur Þór. "Opening Address at the seminar 'Doing Business in the Arctic,'" American-Icelandic Chamber of Commerce, Washington, D.C., May 23, 2019c. https://www.stjornarradid.is/raduneyti/utanrikisraduneytid/utanrikisradherra/fyrri-radherrar/stok-raeda-fyrrum-radherra/2019/05/23/Opnunaravarp-a-malthingi-Islensk-Ameriska-vidskiptaradsins-Doing-Business-in-the-Arctic/.

Monographs and Articles

Ackrén, Maria, and Rasmus Leander Nielsen. "The Fist Foreign- and Security Policy Opinion Poll in Greenland." Ilisimatusarfik/University of Greenland and Konrad Adenauer Stiftung, February 2021. https://uni.gl/media/6762444/fp-survey-2021-ilisimatusarfik.pdf.

Adelman, Kenneth L. *Reagan at Reykjavik: Forty-Eight Hours That Ended the Cold War*. New York: Broadside Books/Harper Collins Publishers, 2014.

Adler, Emanuel, and Michael Barnett. "A framework for the study of security communities." In *Security Communities*, edited by Emanuel Adler and Michael Barnett, 29–65. Cambridge: Cambridge University Press, 1998.

Aalabaf-Sabaghi, Morteza. "Bazaar Governance," *SSRN Electronic Journal*, February 27 (2008): 1–5. https://doi.org/10.2139/ssrn.1098783.

Alexeeva, Olga V., and Frédéric Lasserre. "The Snow Dragon: China's Strategies in the Arctic." *China Perspectives* 3 (2012): 61–68.

Aliber, Robert Z., and Gylfi Zoega, eds. *Preludes to the Icelandic Financial* Crisis. New York: Palgrave Macmillan, 2011.

Anderson, Alun. *After the Ice: Life, Death, and Geopolitics in the New Arctic*. New York: Smithsonian Books, 2009a.

Anderson, D.H. "The Status Under International Law of the Maritime Areas Around Svalbard." *Ocean Development & International Law*, xl, no. 4 (2009b): 373–84.

Andreatta, Filippo, and Mathias Koenig-Archibugi. "Which Synthesis? Strategies of Theoretical Integration and the Neorealist-Neoliberal Debate." *International Political Science Review* 31, no. 2 (2010): 207–227.

Archer, Clive. "The Stoltenberg Report and Nordic security: big idea, small steps," *Danish Foreign Policy Yearbook*, 43–74. Copenhagen: DIIS, 2010.

Ásgeirsdóttir, Áslaug. *Who Gets What? Domestic Influences on International Negotiations Allocating Shared Resources*. New York: Suny Press, 2008a.

Ásgeirsdóttir, Áslaug. "Á hafi úti: Áhrif hagsmunahópa á samninga Íslendinga og Norðmanna vegna veiða úr flökkustofnum" [In Distant Waters: The Influence of Interest Groups on Agreements between Iceland and Norway on Straddling Stocks Fishing Activities]. In *Uppbrot hugmyndakerfis. Endurmótun íslenskrar utanríkisstefnu 1991–2007* [The Unravelling of an Ideational System: Reconfiguring Icelandic Foreign and Security Policy, 1991–2007], edited by Valur Ingimundarson, 349–371. Reykjavík: Hið íslenska bókmenntafélag, 2008b.

Baldwin, David A., ed. *Neo-realism and Neo-liberalism: The Contemporary Debate*. New York: Columbia University Press, 2003.

Bailes, Alyson J.K., and Lassi Heininen. *Strategy Papers on the Arctic or High North: a comparative study and analysis*. Reykjavík: Centre for Small States, 2012.

Bailes, Alyson J.K., and Örvar Þ. Rafnsson. "Iceland and the EU's common and security and defence policy: challenge or opportunity?" *Stjórnmál og stjórnsýsla* 8, no. 1 (2012): 109–131.

Barry, Tom, Brynhildur Davíðsdóttir, Níels Einarsson, and Oran R. Young. "The Arctic Council: an agent of change?" *Global Environmental Change* 63, no. 37 (2020), 102099. https://doi.org/10.1016/j.gloenvcha.2020.102099.

Bennett, Mia. "Rise of Sinocene: China as a Geological Agent." In *Infrastructure and the Remaking of Asia*, edited by Max Hirsh and Till Mostowlansky, 19–41. Honolulu: University of Hawai'i Press, 2020.

Bennett, Mia, Scott R. Stephenson, Kang Yang, Michael T. Bravo, and Bert Det Jonghe. "The opening of the Transpolar Sea Route: Logistical, environmental, and socioeconomic impacts." *Marine Policy*, 121 (2020). https://doi.org/10.1016/j.marpol.2020.104178.

Bergmann, Eiríkur. *Iceland and the International Financial Crisis: Boom, Bust and Recovery*. Basingstoke and New York: Palgrave Macmillan, 2014.

Bergmann, Eiríkur. "The Icesave Dispute: A Case Study into the Crisis of Diplomacy during the Credit Crunch." *Nordicum Mediterraneum*, 12, no. 1 (2017). https://nome.unak.is/wordpress/volume-12-no-1-2017/double-blind-peer-reviewed-article/icesave-dispute-case-study-crisis-diplomacy-credit-crunch/.

Berkmann, Paul Arthur, Alexander N. Vylegzhanin, and Oran R. Young. "Application and interpretation of the Agreement on Enhancing International Arctic Scientific Cooperation." *Moscow Journal of International Law* 49, no. 3 (2017): 6–17.

Beukel, Erik, Frede P. Jensen, and Jens Elo Rytter. *Phasing Out the Colonial Status of Greenland, 1945–55: A Historical Study*. Copenhagen: Museum Tusculanum Press, 2010.

Billig, Michael. *Banal Nationalism*. London: Sage Publications, 1995.

Bittner, Donald F. The Lion and the White Falcon: Britain and Iceland in the World War II Era. Archon Books: Hamden, Connecticut 1983.

Bjarnason, Gunnar Þór. *Óvænt áfall eða fyrirsjáanleg tímamót? Brottför Bandaríkjahers frá Íslandi. Aðdragandi og viðbrögð* [An Unexpected Shock or a Predictable Turning Point? The Withdrawal of U.S. Troops from Iceland: Prehistory and Reaction]. Reykjavík: University of Iceland Press, 2008.

Bloom, Evan T. "Establishment of the Arctic Council." *The American Journal of International Law* 93, no. 3 (July 1993a): 712–722.

Bloom, Lisa. *Gender on the Ice: American Ideologies of Polar Expeditions*. Minneapolis: University of Minnesota Press, 1993b.

Borgerson, Scott G. "Arctic Meltdown: The Economic and Security Implications of Global Warming." *Foreign Affairs* 87, no. 2 (March/April 2008): 63–77.

Borgerson, Scott G. "The Great Game Moves North." *Foreign Affairs*, March 25, 2009. https://www.foreignaffairs.com/articles/commons/2009-03-25/great-game-moves-north.

Borgerson, Scott G. "The Coming of the Arctic Boom: As the Ice Melts, the Region Heats Up." *Foreign Affairs*, 92, no. 4 (2013): 76–89.

Boyes, Roger. *Meltdown Iceland: How the Global Financial Crisis Bankrupted an Entire Country*. London: Bloomsbury, 2009.

Brady, Anne Marie. *China as a Polar Great Power*. Washington, D.C.: Woodrow Wilson Center Press, 2017.

Broek, Emilie, Broek, Nicholaz Olczak, and Lisa Dellmuth. "The Involvement of Civil Society Organizations in Arctic Governance." *SIPRI Insights on Peace and Security*, 2 (2023): 1–28.

Brown, Archie. "The Gorbachev revolution and the end of the Cold War." In *Cambridge History of the Cold War*, edited by Melvyn P. Leffler and Odd Arne Westad, 244–266. Cambridge: Cambridge University Press, 2010.

Buchanan, Elizabeth. *Red Arctic: Russian Strategy Under Putin*. Lanham, Rowman & Littlefield, 2023.

Burke, Danita Catherine. "Leading by example: Canada and its Arctic stewardship role." *International Journal of Public Policy* 13, no. 1–2 (2017): 36–52.

Calderwood, Cayla, and Frances Ann Ulmer. "The Central Arctic Ocean fisheries moratorium: A rare example of the precautionary principle in fisheries management." *Polar Record* 59 (2023), (e1), January 16, 2023: 1–14. https://doi.org/10.1017/S0032247422000389.

Cambo, Dorothée Céline. "Disentangling the conundrum of self-determination and its implications in Greenland." *Polar Record* 56 (e3): 1–10. https://doi.org/10.1017/S0032247420000169.

Campion, Andrew Stephen. "From CNOOC to Huawei: securitization, the China Threat, and critical infrastructure." *Asian Journal of Political Science*, 29, no 1 (2020): 47–66.

Cela, Margrét. "Iceland: A Small Arctic State Facing Big Arctic Changes." *The Yearbook of Polar Law* V (2013): 75–92.

Cela, Margrét, and Pia Hansson. "A challenging chairmanship in turbulent times." *The Polar Journal* 11, no. 1 (2021): 43–56.

Ciorciari, John D. "The variable effectiveness of hedging strategies." *International Relations of the Asia-Pacific* 19, no. 3 (2019): 523–555.

Ciorciari, John D., and Jürgen Haacke. "Hedging in international relations: an introduction." *International Relations of the Asia-Pacific* 19, no. 3 (2019): 367–374.

Chamberlain, Muriel Evelyn. *The Scramble for Africa*. London, New York: Routledge, 2013.

Chater, Andrew. "Explaining Non-Arctic States in the Arctic Council." *Strategic Analysis* 40, no. 3 (2016): 173–184.

Chatterjee, Partha. *The Nation and Its Fragments: Colonial and Postcolonial Histories*. Princeton: Princeton University Press, 1993.

Churchill, Robin, and Geir Ulfstein. "The Disputed Maritime Zones around Svalbard." In *Changes in the Arctic Environmental and the Law of the Sea*, edited by Myron H. Nordquist, Tomas H. Heidar, and John Norton Moore, 551–593. Leiden: Martinus Nijhoff, 2010.

Corgan, Michael T. "Franklin D. Roosevelt and the American Occupation of Iceland." *Naval War College Review*, 45, no. 4 (1992): 34–54.

Corgan, Michael T. *Iceland and Its Alliances: Security for a Small State*. Lewiston, N.Y.: E. Mellen Press, 2002.

Crompton, Peter. "Hedging in academic writing: some theoretical problems." *English for Specific Purposes* 16, no. 4 (1997): 271–287.

De Carvalho, Benjamin, and Ivar Neumann. "Small states and status." In *Small state status seeking. Norway's Quest for International Standing*, edited by Neumann and De Carvalho, 56–72. London: Routledge, 2015.

Depledge, Duncan. "NATO and the Arctic: The Need for a New Approach." *The RUSI Journal* 165, no. 5–6 (2020): 80–90.

Depledge, Duncan, and Klaus Dodds, "Bazaar Governance: Situating the Arctic Circle." In *Governing Arctic Change: Global Perspectives*," edited by Kathrin Keil and Sebastian Knecht, 141–160. London: Palgrave Macmillan, 2017.

Deutsch, Karl. *Political Community and the North Atlantic Area*. Princeton: Princeton University Press, 1957.

Devyatki, Pavel. "Arctic exceptionalism: a narrative of cooperation and conflict from Gorbachev to Medvedev and Putin," *Polar Journal* 13, no. 2 (2023): 336–357.

Dittmer, Jason, Sami Moisio, Alan Ingram, and Klaus Dodds. "Have you heard the one about the disappearing ice? Recasting Arctic geopolitics." *Political Geography* 30, no. 4 (2011): 202–214.

Dodds, Klaus, and Valur Ingimundarson. "Territorial nationalism and Arctic geopolitics: Iceland as an Arctic coastal state." *The Polar Journal* 2, no. 1 (2012): 21–37.

Dodds, Klaus. "The Ilulissat Declaration (2008): The Arctic States, 'Law of the Sea,' and Arctic Ocean." *SAIS Review of International Affairs* 33, no. 2 (2013): 45–55.

Dukes, Paul. "Vilhjalmur Stefansson: The Northward Course of Empire, The Adventure of Wrangel Island, 1922-1925, and 'Universal Revolution'." *Sibirica: Interdisciplinary Journal of Siberian Studies* 17, no. 1 (2018), 1–22.

Duque, Marina G. "Recognizing International Status: A Relational Approach." *International Studies Quarterly* 62, no. 3 (2018): 577–592.

Durrenberger, E. Paul, and Gisli Palsson, eds. *Gambling Debt: Iceland's Rise and Fall in the Global Economy*. Boulder: University of Colorado Press, 2014.

Edmonds, David, and John Eidinow, *Bobby Fischer Goes to War*. New York: Harper Publishers, 2004.

Eilstrup-Sangiovanni, Mette. "Uneven Power and the Pursuit of Peace: How Regional Power Transitions Motive Integration." *Comparative European Politics* 6 (2008): 102–142.

Einarsdóttir, Vilborg. "Ólafur Ragnar Grímsson: A New Model of Arctic Cooperation for the 21st Century." *JONAA, Journal of the North Atlantic &*

Arctic (November 2018). https://www.jonaa.org/content/2018/10/19/a-new-model.

Einarsson, Sveinn K., Ingjaldur Hannibalsson, and Alyson J.K. Bailes. "Chinese Investment and Icelandic National Security." In *Þjóðarspegillinn*, edited by Silja B. Ómarsdóttir, 1–13. Reykjavík Félagsvísindastofnun Háskóla Íslands, 2014. https://www.semanticscholar.org/paper/Chinese-Investment-and-Icelandic-National-Security-Einarsson-Bailes/6251bacb83a9c6b6e24bd7b84b29a1d5973fe42c.

Eriksen, Erik Oddvar, and John Erik Fossum, eds. *The European Union's non-members: independence under hegemony?* London and New York, 2015.

Exner-Pirot, Heather. "Between Militarization and Disarmament: Challenges for Arctic Security in the Twenty-First Century." In *Climate Change and Arctic Security: Searching for a Paradigm Shift*, edited by Lassi Heininen and Heather Exner-Pirot, 91–206. Cham: Palgrave/Macmillan, Springer Nature: 2020.

Exner-Pirot, Heather, and Robert Murray. *Regional Order in the Arctic: Negotiated Exceptionalism*. The Arctic Institute, October 24, 2017, 47–63. https://www.thearcticinstitute.org/regional-order-arctic-negotiated-exceptionalism/.

Farnham, Barbara. "Reagan and the Gorbachev Revolution: Perceiving the End of Threat." *Political Science Quarterly* 116, no. 2 (2001): 225–252.

Filtenborg, Mette Sicard, Stefan Gänzle, and Elisabeth Johansson. "An Alternative Theoretical Approach to EU Foreign Policy: 'Network Governance' and the Case of the Northern Dimension." *Cooperation and Conflict* 37, no. 4 (2003): 387–407.

Finger, Matthias, and Gunnar Rekvig. *Global Arctic: An Introduction to the Multifaceted Dynamics of the Arctic*. Cham: Springer, 2022

Foggo, James G., and Alarik Fritz. "NATO and the Challenge in the North Atlantic and the Arctic." In *Northern Europe: Deterrence, Defence and Dialogue*, edited by John Andreas Olsen, 121–128. Whitehall Papers. London: RUSI, 2018.

Fondahl, Gail, Aileen A. Espiritu, and Aytalina Ivanova. "Russia's Arctic Regions and Policies." In *The Palgrave Handbook of Arctic Policy and Politics*, edited by Ken S. Coates and Carin Holroyd, 195–216. London: Palgrave Macmillan, 2019.

Foot, Rosemary. "Chinese strategies in a US-hegemonic global order: accommodating and hedging." *International Affairs* 82, no. 1 (2006): 77–94.

Forsberg, Tuomas. "The rise of Nordic defense cooperation: a return to regionalism?" *International Affairs* 89, no. 5 (2013): 1161–1181.

Friðriksson, Guðjón. *Saga af forseta* [A President's Account]. Reykjavík: Forlagið, 2008.

Garthoff, Raymond L. *The Great Transition: American-Soviet Relations and the End of the Cold War*. Washington, D.C.: Brookings Institution, 1994.

Grant, Shelagh D. *Polar Imperative: A History of Arctic Sovereignty in North America*. Vancouver: Douglas and Mclntyre, 2010.

Graczyk, Piotr. "Observers in the Arctic Council – Evolution and Prospects." *Yearbook of Polar Law* 3 (2011): 594–596.

Graczyk, Piotr, and Svein Vigeland Rottem, "The Arctic Council: soft actions, hard effects?" In *Routledge Handbook of Arctic Security,* edited by Gunnhild Hoogensen Gjörv, Marc Lanteigne, and Horatio Sam-Aggrey, 221–233. London and New York: Routledge, 2020.

Grajewski, Nicole Bayat. "Russia's Great Power Assertion: Status-Seeking in the Arctic." *St. Anthony's International Review*, 13, no. 1 (2017): 141–163.

Greaves, Wilfried, and P. Whitney Lackenbauer, eds. *Breaking Through: Understanding Sovereignty in the Circumpolar Arctic.* Toronto: University of Toronto Press, 2021.

Gricius, Gabriella, and Erin B. Fitz. "Can Exceptionalism Withstand Crises? An Evaluation of the Arctic Council's Response to Climate Change and Russia's War on Ukraine." *Global Studies Quarterly* 2, no. 3 (2022): 1–6.

Griffiths, Franklyn. "Towards a Canadian Arctic Strategy." *Zeitschrift für öffentliches Recht und Völkerrecht* (*ZaöRV*), 69 (2009): 579–624.

Griffiths, Franklyn. *The Politics of the Northwest Passage.* Montreal: McGill-Queens University Press, 1987.

Grímsson, Ólafur Ragnar, *Sögur handa Kára* [Stories for Kári [Stefánsson]]. Reykjavík: Mál og menning, 2020)

Guðmundsson, Guðmundur J. "The Cod and the Cold War." *Scandinavian Journal of History* 31, no. 2 (2006): 97–118.

Gudmundsson, Thorir. "Cod War on the High Seas: Norwegian-Icelandic dispute over the 'Loophole' fishing in the Barents Sea." *Nordic Journal of International Law* 64 (1995): 557–572.

Gunnarsson, Gunnarsson. "Continuity and Change in Icelandic Security and Foreign Policy." *The Annals of American Academy of Political and Social Science*, 512, no. 1 (1990): 140–151

Gunnarsson, Styrmir. *Umsátrið – fall Íslands og endurreisn* [Under Siege: The Collapse of Iceland and Its Reconstruction]. Reykjavík: Veröld 2009.

Gunnarsson, Þorsteinn, and Egill Þór Níelsson. "An Icelandic perspective." In *Nordic-China Cooperation. Challenges and Opportunities*, edited by Andreas Bøje Forsby, 87–93. Copenhagen: NIAS Press, 2019.

Heininen, Lassi. Heather Exner-Pirot, and Justin Barnes. "'Introduction': Redefining Arctic security: Military, environmental, human or societal? Cooperation or conflict?" In *Arctic Yearbook 2019*, edited by Lassi Heininen, Heather Exner-Pirot, and Justin Barnes, 1–8. Akureyri: Arctic Portal, 2019.

Hobsbawm, Eric J. *Nations and Nationalism since 1780.* Cambridge: Cambridge University Press, 1990.

Haacke, Jürgen. "The concept of hedging and its application to Southeast Asia: a critique and a proposal for a modified conceptual and methodological framework." *International Relations of the Asia-Pacific* 19, no. 3 (2019): 375–417.

Hacquebord, Louwrens. "How Science Organizations in the Non-Arctic Countries became Members of IASC." *IASC after 25 Years*, Special Issue of the *IASC Bulletin* (2015): 21–26.

Hálfdanarson, Guðmundur. "'The Beloved War' The Second World War and the Icelandic National Narrative." In *Nordic Narratives of the Second World War: National Historiographies Revisited*, edited by Henrik Stenius, Mirja Österberg, and Johan Östling. Lund: Nordic Academic Press, 2011.

Hansson, Pia Elísabet, and Guðbjörg Ríkey Th. Hauksdóttir. "Iceland and Arctic Security: US Dependency and the Search for an Arctic Identity." In *On Thin Ice? Perspectives on Arctic Security*, edited by Duncan Depledge and P. Whitney Lackenbauer, 163–171. Ontario: North American and Arctic Defense and Security Network (NAADSN), 2021.

Haugvik, Kristin and Ulf Sverdrup, eds. "Ten Years On: Reassessing the Stoltenberg Report on Nordic Cooperation." NUPI, IIA, FIIA, DIIS, UI: Oslo, Helsinki, Copenhagen, Stockholm, and Reykjavík, 2019.

Hauksson, Daníel Hólmar. "Grænlandsdraumurinn. Hugmyndir um tilkall Íslendinga til Grænlands á 20. öld" [The Greenland Dream: Ideas about Icelandic Claims to Greenland]. BA thesis, University of Iceland, 2019.

Heininen, Lassi, Karen Everett, Barbora Padrtova, and Anni Reissell. *Arctic policies and strategies – analysis, synthesis, and trends.* Laxenburg, Austria: International Institute for Applied Systems Analysis, 2020.

Heininen, Lassi, and Heather N. Nicol. "The Importance of Northern Dimension Foreign Policies in the Geopolitics of the Circumpolar North." *Geopolitics* 12, no. 1 (2007): 133–165.

Heininen, Lassi. "The Northern Research Forum – a Pioneering Model for an Open Discussion." *Arctic Circle Journal*, September 14, 2023. https://www.arcticcircle.org/journal/the-northern-research-forum.

Hellmann, Gunther, and Benjamin Herborth. "Fishing in the mild West: democratic peace and militarised interstate disputes in the transatlantic community." *Review of International Studies* 34, no. 3 (2008): 481–506.

Hirsh, Max, and Till Mostowlansky, eds. *Infrastructure and the Remaking of Asia.* Honolulu University of Hawai'i Press, 2020.

Holtsmark, Sven G. and Brooke A. Smith-Windsor, eds. *Security prospects in the High North: geostrategic thaw or freeze?* Rome: NATO Defense College, 2009. https://www.isn.ethz.ch/Digital-Library/Publications/Detail/?id=102391&lng=en.

Hong, Nong. "Emerging interests of non-Arctic countries in the Arctic: A Chinese perspective." *The Polar Journal* 4, no. 2 (2014): 271–286.

Hong, Nong. *China's Role in the Arctic: Observing and Being Observed.* London, New York: Routledge, 2020.

Hunt, Jonathan, and David Reynolds. "Geneva, Reykjavik, Washington, and Moscow, 1985–8." In *Transcending the Cold War: Summits, Statecraft, and the*

Dissolution of Bipolarity in Europe, 1970–1990, edited by Kristina Spohr and David Reynolds, 151–79. Oxford: Oxford University Press, 2016.

Hønneland, Geir. *Arctic Euphoria and International High North Politics*. Singapore: Springer, 2017.

Ingimarsson, Skafti. "Íslenskir kommúnistar og sósíalistar: Flokksstarf, félagsgerð og stjórnmálabarátta 1918–1968" [Icelandic Communists and Socialists: Party Activism, Social Structure, and Political Struggle]. Ph.D. diss. University of Iceland, 2018.

Ingimundarson, Valur. *Í eldlínu kalda stríðsins. Samskipti Íslands og Bandaríkjanna 1945–1960* [In the Crossfire: Icelandic-U.S. Relations 1945–1960]. Reykjavík: Vaka Helgafell, 1996.

Ingimundarson, Valur. *Uppgjör við umheiminn: Samskipti Íslands við Bandaríkin og NATO 1960–1974* [A Reckoning with the Outside World: Iceland's Relations with the United States and NATO, 1960–1974]. Reykjavík: Vaka Helgafell, 2001a.

Ingimundarson, Valur. "Buttressing the West in the North: The Atlantic Alliance, Economic Warfare, and the Soviet Challenge in Iceland, 1956–1959." *The International History Review*, 21, no. 1 (1999): 80–103.

Ingimundarson, Valur. "Icelandic Domestic Politics and Popular Perceptions of NATO, 1949–1999." In *NATO—The First Fifty Years*, ed. Gustav Schmidt, 285–302. London: Macmillan, 2001b.

Ingimundarson, Valur. "Confronting Strategic Irrelevance: The End of a US-Icelandic Security Community?" *RUSI Journal* 150, no. 4 (2005): 66–71.

Ingimundarson, Valur. "In memoriam: Orðræða um orrustuþotur, 1961–2006" [In Memoriam: Discourse on U.S. Fighter Jets in Iceland], *Skírnir* 180, no. 2 (summer) 2006: 31–60.

Ingimundarson, Valur. "Frá óvissu til upplausnar: 'Öryggissamfélag' Íslands og Bandaríkjanna 1991–2006" [From Uncertainty to Dissolution: The Icelandic--U.S. Security Community, 1991–2008]. In *Uppbrot hugmyndakerfis. Endurmótun íslenskrar utanríkisstefnu 1991–2007* [The Unravelling of an Ideational System: Reconfiguring Icelandic Foreign and Security Policy, 1991–2007], edited by Ingimundarson, 1–66. Reykjavík: Hið íslenska bókmenntafélag, 2008.

Ingimundarson, Valur. "Iceland's Post-American Security Policy, Russian Geopolitics and the Arctic Question." *RUSI Journal* 154, no. 4 (2009): 74–80.

Ingimundarson, Valur. "'A Crisis of Affluence': the Politics of an Economic Breakdown in Iceland." *Irish Studies in International Affairs* 21 (2010): 57–69.

Ingimundarson, Valur. *The Rebellious Ally: Iceland, the United States, and the Politics of Empire, 1945–2006*. Dordrecht and St. Louis: Republic of Letters Publishing, 2011a.

Ingimundarson, Valur. "Territorial Discourses and Identity Politics: Iceland's Role in the Arctic." In *Arctic Security in an Age of Climate Change*, edited by James Kraska, 174–189. Cambridge: Cambridge University Press, 2011b.

Ingimundarson, Valur. "Managing a contested region: the Arctic Council and the politics of Arctic governance," *The Polar Journal* 4, no. 1 (2014): 183–198.

Ingimundarson, Valur. "Framing the national interest: the political uses of the Arctic in Icelandic foreign and domestic policies." *The Polar Journal* 5, no. 1 (2015): 81–100.

Ingimundarson, Valur, and Halla Gunnarsdóttir. "The Icelandic Sea Areas and Activity Level up to 2025." In *Maritime activity in the High North – current and estimated level up to 2025*, edited by Odd Jarl Borch et al., 74–85. Bodø: Nord University, 2016.

Ingimundarson, Valur, Philippe Urfalino, and Irma Erlingsdóttir, eds. *Iceland's Financial Crisis: The Politics of Blame, Protest, and Reconstruction*. New York and London: Routledge, 2016.

Ingimundarson, Valur. "A Fleeting or Permanent Military Presence? The Revival of US Anti-Submarine Operations from Iceland." *RUSI Newsbrief* 38, no. 7 (2018a): 1–4.

Ingimundarson, Valur. "The Geopolitics of the 'Future Return': Britain's Century-Long Challenges to Norway's Control over Spitsbergen." *The International History Review* 40, no. 4 (2018b): 893–915.

Ingimundarson, Valur. "Iceland as an Arctic State." In *The Palgrave Handbook of Arctic Policy and Politics*, edited by Ken S. Coates and Carin Holroyd, 251–265. London: Palgrave Macmillan, 2019.

Ingimundarson, Valur. "Unarmed sovereignty versus foreign base rights: enforcing the US-Icelandic defence agreement 1951–2021." *The International History Review* 44, no. 1 (2022): 73–91.

Ingimundarson, Valur. "Interpreting Iceland's victories in the 'Cod Wars' with the United Kingdom." In *The Success of Small States in International Relations: Mice that Roar?* edited by Godfrey Baldacchino, 37–50. London and New York: Routledge, 2023.

Ísleifsson, Sumarliði R. *Í fjarska norðursins. Ísland og Grænland – viðhorfasaga í þúsund ár* [In the Distant North: Iceland and Greenland—a Thousand-Year History of External Images]. Reykjavík: Sögufélagið, 2020.

Ísleifsson, Sumarliði R. with the collaboration of Daniel Chartier: *Iceland and Images of the North*. Québec: Presses de L'université de Québec, 2011.

Jakobson, Linda, and Jingchao Peng. *China's Arctic Aspiration*. SIPRI Policy Paper no. 34 Stockholm: International Peace Research Institute, 2012.

Jensdóttir Harðarson, Sólrún B. "The 'Republic of Iceland' 1940–44: Anglo-American Attitudes and Influences." *Journal of Contemporary History*, 9, no. 4 (1974): 27–56.

Jervis, Robert. "Realism, Neoliberalism, and Cooperation: Understanding the Debate." *International Security*, 24, no. 1 (1999): 42–63.

Johannsdottir, Lara, and David Cook. "Discourse analysis of the 2013–2016 Arctic Circle Assembly programmes." *Polar Record* 53, no. 3 (2017): 276–279.

Jóhannesson, Guðni Th. "How Cod War Came: The Origins of the Anglo-Icelandic Fishing Dispute, 1958–61." *Historical Research*, 77, no. 198 (2004a): 543–574.

Jóhannesson, Guðni Th. "To the Edge of Nowhere?" *Naval War College Review* 57, no. 3 (2004b): 115–137.

Jóhannesson, Guðni Th. *Troubled Waters: Cod War, Fishing Disputes, and Britain's Fight for the Freedom of the High Seas, 1948–1964.* Reykjavík: North Atlantic Fisheries History Association, 2007.

Jóhannesson, Guðni Th. *Hrunið* [The Crash]. Reykjavík: JPV, 2009.

Johnsen, Guðrún. *Bringing Down the Banking System: Lessons from Iceland.* New York: Palgrave Macmillan, 2014.

Johnston, Alastair Iain, and Robert S. Ross, eds. *Engaging China: The Management of an Emerging Power.* London and New York: Routledge, 1999.

Jónasson, Ögmundur. *Rauði þráðurinn* [The Common Thread]. Selfoss: Bókaútgáfan Sæmundur, 2022.

Jónsdóttir, Jóhanna. *Iceland and the EU: Europeanization and the European Economic Area.* Abington, Oxon, New York: Routledge, 2013.

Jónsson, Albert. *Iceland, NATO, and the Keflavik Base.* Reykjavík: Icelandic Commission on Security and International Affairs, 1989a.

Jónsson, Ásgeir. *Why Iceland? How One of the World's Smallest Countries Became the Meltdown's Biggest Casualty.* New York: McGraw-Hill, 2009.

Jónsson, Ásgeir, and Hersir Sigurgeirsson. *The Icelandic Financial Crisis: A Study into the World's Smallest Currency Area and its Recovery from Total Banking Collapse.* London: Palgrave, 2016.

Jónsson, Hannes. "Íslensk hlutleysisstefna [Iceland's Neutrality Policy]." *Andvari* 114, no. 1 (1989b): 203–224.

Jónsson, Örn Daníel, Ingjaldur Hannibalsson, and Li Yang. "A bilateral free trade agreement between Iceland and China." In *Þjóðarspegillinn – Rannsóknir í félagsvísindum*, edited by Ingjaldur Hannibalsson, 1–7. Reykjavík: Félagsvísindastofnun Háskóla Íslands, 2013. https://skemman.is/en/stream/get/1946/16786/39049/3/OrnDJonsson_VID.pdf.

Lackenbauer, P. Whitney, and Suzanne Lalonde, eds. *Breaking the Ice Curtain? Russia, Canada, and Arctic Security in a Changing Circumpolar World.* Calgary: Canadian Global Affairs Institute, 2019.

Lagutina, Maria L. *Russia's Arctic Policy in the Twenty-first Century: National and International Dimensions.* Lanham: Lexington Books, 2019.

Lake, David. *Hierarchy in International Relations.* Ithaca: Cornell University Press, 2009.

Lanteigne, Marc. "The Rise (and Fall?) of the Polar Silk Road." *The Diplomat*, August 29, 2022. https://thediplomat.com/2022/08/the-rise-and-fall-of-the-polar-silk-road/.

Lasserre, Frédéric, Jérôme Le Roy, and Richard Garon. "Is There an Arms Race in the Arctic." *Centre of Military and Strategic Studies*, 14, 3–4 (2012): 1–56.

Lavelle, Kathryn C. "Regime, Climate, and Region in Transition: Russian Participation in the Arctic Council." *Problems of Post-Communism* 69, no. 4–5 (2022): 345–357.

Lynch, Allen C. "The influence of regime type on Russian foreign policy toward 'the West,' 1992–2015." *Communist and Post-Communist Studies* 49, no. 1 (2016): 101–111.

Lynch, Allen C. "The Realism of Russia's Foreign Policy." *Europe-Asia Studies*, 53, no. 1 (2001): 7–31.

Lim, Darren J. "Hedging in South Asia: balancing economic and security interests amid Sino-Indian competition." *International Relations of the Asia-Pacific* 19, no. 3 (2019), 493–522.

Loftsdóttir, Kristín. "Racist Caricatures in Iceland in the early 20th century." In *Iceland and Images of the North*, edited by Sumarliði R. Ísleifsson with the collaboration of Daniel Chartier, 187–204. Québec: Presses de L'université de Québec, 2011.

Loftsdóttir, Kristín. "The Exotic North: Gender, Nation Branding and Nationalism in Iceland." *Nora – Nordic Journal of Feminist and Gender Research*, 23, No. 4 (2015): 246–260.

Käpylä, Juha, and Harri Mikkola. "On Arctic Exceptionalism." The Finnish Institute of International Affairs, Working Paper 85 (April 2015), 1–22. https://www.fiia.fi/wp-content/uploads/2017/01/wp85.pdf

Kankaanpää, Pauka, and Oran R. Young. "The effectiveness of the Arctic Council." *Polar Research* 31, no. 1 (2012): 1–14.

Kegley, Charles. *Controversies in International Relations Theory: Realism and the Neoliberal Challenge*. New York: St. Martin's Press, 2015.

Kjellén, Jonas. "The Russian Northern Fleet and the (Re) militarization of the Arctic." *Arctic Review on Law and Politics*, 13 (2022): 34–52.

Keohane, Robert O. *International Institutions and State Power: Essays in International Relations Theory*. New York: Routledge, 2020.

Keohane, Robert O., ed. *Neorealism and Its Critics*. New York: Columbia University Press, 1986.

Keohane, Robert O., and Joseph J. Nye. *Power and Interdependence*. 3rd ed. London: Longman, 2000.

Kikkert, Peter, and P. Whitney Lackenbauer. "The Militarization of the Arctic to 1990." In *The Palgrave Handbook of Arctic Policy and Politics*, edited by Ken S. Coates and Carin Holroyd, 487–505. Cham: Palgrave Macmillan, 2020.

Kobzeva, Mariia. "Strategic partnership setting for Sino-Russian cooperation in Arctic Shipping." *The Polar Journal 10*, no. 2 (2020): 334–352.

Koivurova, Timo. "The Arctic Council at 10 Years: Retrospect and Prospects." *University of British Columbia Law Review* 40, no. 1 (2007): 121–194.

Koivurova, Timo, and Akiho Shibata. "After Russia's invasion of Ukraine in 2022: Can we still cooperate with Russia in the Arctic." *Polar Record*, 59 (2023) (e12), March 17, 2023: 1–9. https://doi.org/10.1017/S0032247423000049.

Korolev, Alexander. "Shrinking room for hedging: system-unit dynamics and behavior of smaller powers." *International Relations of the Asia-Pacific* 19, no. 3 (2019): 419–452.

Korolev, Alexander. "Systemic Balancing and Regional Hedging: China-Russia Relations." *Chinese Journal of International Politics* 9, no. 4 (2016): 375–397.

Kraska, James. "International Security and International Law in the Northwest Passage." *Vanderbilt Law Review* 42, no. 4 (2009): 1109–1132.

Kristjánsdóttir, Helga, Sigurður Guðjónsson, and Guðmundur Kristján Óskarsson. "Free Trade Agreement (FTA) with China and Interaction between Exports and Imports," *Baltic Journal of Economic Studies* 8, no. 1 (2022), DOI: https://doi.org/10.30525/2256-0742/2022-8-1-1-8.

Kuik, Cheng-Chwee. "How Do Weaker States Hedge? Unpacking ASEAN States' alignment behavior towards China." *Journal of Contemporary China* 25, no. 100 (2016): 500–514.

Kuik, Cheng-Chwee. "Getting hedging right: a small state perspective." *China International Strategy Review* 3, no. 2 (2021): 300–315.

Kuokkanen, Rauna. "'To See What State We Are In': First Years of the *Greenland Self-Government Act* and the Pursuit of Inuit Sovereignty." *Ethnopolitics* 16, no. 2 (2017): 179– 195.

Kurlantzik, Joshua. *Charm Offensive: How China's Soft Power is Transforming the World*. New Haven: Yale University Press, 2007.

Maas, Matthias. "The elusive definition of the small state." *International Politics* 46, no. 1 (2009): 65–83.

Macaraig, Christine Elizabeth, and Adam James Fenton. "Analyzing the Causes and Effects of the South China Sea Dispute." *The Journal of Territorial and Maritime Studies*, 8, no. 2 (2021): 42–58.

Magnúss, Gunnar M. *Virkið í norðri* [The Fortress in the North], I–III. Reykjavík: Bókaútgáfan Virkið, 1984.

Magnússon, Bjarni Már. "The Loophole Dispute from an Icelandic Perspective." Working paper, 1–31. Centre for Small States, University of Iceland, 2010.

Magnússon, Bjarni Már. "Nokkrir farvegir fyrir hafréttardeilur Íslands við erlend ríki og alþjóðastofnanir [Several Options in Iceland's Law of the Sea Disputes between Iceland and Other States and International Organizations]." *Tímarit Lögréttu* 10, no. 1 (2014): 9–20.

Magnússon, Bjarni Már. *The Continental Shelf Beyond 200 Nautical Miles: Delineation, Delimitation and Dispute Settlement.* Leiden and Boston: Brill Martinus Nijhoff, 2015.

Magnússon, Bjarni Már, and Charles H. Norchi. "Geopolitics and International Law in the Arctic." In *Routledge Handbook of Arctic Security*, edited by Gunhild Hoogensen Gjørv, Marc Lanteigne, and Horatio Sam-Aggrey, 246–257. London and New York: Routledge/Taylor & Francis Group, 2020.

Marchenko, Nataliya, Odd Jarl Borch, Natalia Andreassen, Svetlana Kuznetsova, Valur Ingimundarson, and Uffe Jakobsen. "Navigation Safety and Risk Assessment Challenges in the High North." In *Marine Navigation and Safety of Sea Transportation*, edited by Adam Weintritt, 275–281. Boca Raton, FL: Taylor and Francis Group, 2017.

Marsh, Rosalind. "The Nature of Russia's Identity: The Theme of 'Russia and the West' in Post-Soviet Culture." *Nationalities Papers* 35, no. 3 (2007): 555–578.

Matlock, Jack F. *Reagan and Gorbachev: How the Cold War Ended.* New York: Random House, 2004.

McCannon, John. *Red Arctic: Polar Exploration and the Myth of the North in the Soviet Union, 1932–1939.* Oxford: Oxford University Press, 1998.

Mearsheimer, John. "Structural Realism." In *International Relations Theories: Discipline and Diversity*, edited by Tim Dunne, Milja Kurki, and Steve Smith, 51–68. Oxford: Oxford University Press, 2007.

Medby, Ingrid A. "Articulating state identity: 'Peopling' the Arctic state." *Political Geography* 62, no. 1 (2018): 116–125.

Molenaar, Erik J. "International Regulation of Central Arctic Ocean Fisheries." In *Challenges of the Changing Arctic: Continental Shelf, Navigation, and Fisheries*, edited by Myron H. Nordquist, John Norton Moore, and Ronán Long, 429–463. Leiden: Brill Nijhoff, 2016.

Molenaar, Eric J. "Participation in the Arctic Ocean Fisheries Agreement. In *Emerging Arctic Legal Orders*," edited by Akiho Shibata, Leilei Zou, Nikolas Sellheim, and Marzia Scopelli, 132–170. Abingdon and New York, Routledge, 2019.

Molenaar, Erik J. "The CAOF Agreement: Key Issues of International Fisheries Law." In *New Knowledge and Changing Circumstances in the Law of the Sea*, edited by Tomas Heidar, 446-476. Leiden/Boston: Brill Nijhoff, 2020.

Morshita, Joji. "The Arctic Five-plus-Five process on central Arctic Ocean fisheries negotiations: Reflecting the interests of Arctic and non-Arctic actors" In *Emerging Legal Orders in the Arctic: The Role of Non-Arctic Actors*, edited by Akiho Shibata, Leilei Zou, Nikolas Sellheim, and Marzia Scopelliti, 109–131. Abington, Oxon: Routledge, 2019.

Murphy, Ann Marie. "Great Power Rivalries, Domestic Politics and Southeast Asian Foreign Policy: Exploring the Linkages." *Asian Security* 13, no. 3 (2017): 165–182.

Nielsen, Kristian H. "Transforming Greenland: Imperial Formations in the Cold War." *New Global Studies* 7, no. 2 (2013): 129–154.
Níelsson, Egill Þór. *The West Nordic Council in the Global Arctic.* Reykjavík: Institute of International Affairs, 2013.
Níelsson, Egill Þór. "China Nordic Arctic Research Center." In *Nordic-China Cooperation. Challenges and Opportunities*, edited by Andreas Bøje Forsby, 59–63. Copenhagen: NIAS Press, 2019.
Níelsson, Egill Þór, and Bjarni Már Magnússon. "China's Arctic policy white paper and its influence on the future of Arctic legal developments." In *Emerging Legal Orders in the Arctic: The Role of Non-Arctic Actors*, edited by Akiho Shibata, Leilei Zou, Nikolas Sellheim, and Marzia Scopelliti, 49–65. Abington, Oxon: Routledge, 2019.
Níelsson, Egill Þór, and Guðbjörg Ríkey Th. Hauksdóttir. "Kina, investeringer og sikkerhetspolitikk: Politikk og perspektiver i Norden – Island" [China, Investments and Security Policy: Politics and Perspectives in the Nordics – Iceland]. *Internasjonal Politikk* 78, no. 1 (2020): 68–78.
Niven, Jennifer. *The Ice Master: The Doomed 1913 Voyage of the Karluk and the Miraculous Rescue of Her Survivors.* New York: Hyperion, 2000.
Nordenmann, Magnus. "Back to the Gap." *The RUSI Journal*, 162, no. 1 (2017): 24–30.
North, Douglas C. *The Arctic Council: Governance within the Far North.* London, New York: Routledge, 2016.
Olesen, Mikkel Runge. "The end of Arctic exceptionalism? A review of the academic debates and what the Arctic prospects mean for the Kingdom of Denmark." *Danish Foreign Policy Review 2020.* Copenhagen: Danish Institute for International Studies, 103–127.
Olesen, Thorsten Borring. "Between Facts and Fiction: Greenland and the Question of Sovereignty." *New Global Studies*, 7, no. 2 (2013): 117–128.
Olsen, John Andreas, ed. *NATO and the North Atlantic: Revitalizing Collective Defence*, Whitehall Papers, Vol. 87. London: RUSI, 2016.
Ólafsson, Jón. *Kæru félagar. Íslenskir sósíalistar og Sovétríkin 1920–1960* [Dear Comrades: Icelandic Socialists and the Soviet Union, 1920–1969]. Reykjavík: Mál og menning, 1999.
Pálsson, Gísli. *Travelling Passions: The Hidden Life of Vilhjalmur* Stefansson. Winnipeg: University of Manitoba Press, 2005.
Paul, T.V., Deborah Welch Larson, and William C. Wohlforth, eds. *Status in World Politics.* Cambridge: Cambridge University Press, 2014.
Pay, Vahid Nick, and Harry Gray Calvo. "Arctic Diplomacy: A Theoretical Evaluation of Russian Foreign Policy in the High North." *Russian Politics* 5, no. 1 (2020): 105–130.

Pedersen, Rasmus Brun. "Bandwagon for Status: Changing Patterns in Nordic States Status-seeking Strategies." *International Peacekeeping* 25, no. 2 (2018): 217–241.

Pedersen, Torbjørn. "The Svalbard Continental Shelf Controversy: Legal Disputes and Political Rivalries." *Ocean Development & International Law,* xxxvii, no. 3–4 (2006): 339–358.

Pedersen, Torbjørn, and Tore Henriksen. "Svalbard's Maritime Zones: The End of Legal Uncertainty?" *The International Journal of Marine and Coastal Law*, xxiv, no. 3 (2009): 141–161.

Pedersen, Torbjørn. "Debates over the Role of the Arctic Council." *Ocean Development & International Law*, 43, no. 2 (2012): 146–156.

Petersen, Nikolaj. "The Arctic as a New Arena for Danish Foreign Policy: The Ilulissat Initiative and Its Implications." In *Danish Foreign Policy Yearbook 2009*, edited by Nanna Hvidt and Hans Mouritzen, 35–78. Copenhagen: DIIS – Danish Institute for International Studies, 2009.

Pétursson, Gustav. "The Defence Relationship of Iceland and the United States and the Closure of the Keflavík Base." PhD. diss., University of Lapland, 2020.

Pharand, Donat. "The Arctic Waters and the Northwest Passage: A Final Revisit." *Ocean Development and International Law* 38 (2007): 3–69.

Pharand, Donat. "Canada's Sovereignty Over the Northwest Passage." *Michigan Journal of International Law*, 10, no. 2 (1989): 653–678.

Pharand, Donat. "Draft Arctic Treaty: An Arctic Region Council" [reprint]. *Northern Perspectives* 19, no. 2 (1991): n.p.

Pharand, Donat. "The Case for an Arctic Region Council and a Treaty Proposal." *Revue generale de droit* 2, 23 (1992): 163–195.

Pincus, Rebecca. "Towards a new Arctic: Changing Strategic Geography in the GIUK Gap." *The RUSI Journal*, 165, no. 3 (2020): 50–58.

Puranen, Email Matti and Sanna Kopra, "China's Arctic Strategy – a Comprehensive Approach in Times of Great Power Rivalry," *Scandinavian Journal of Military Studies*, 6, no 1 (2023): 239–253.

Purver, Ronald, "Arctic Security: The Murmansk Initiative." In *Soviet Foreign Policy: New Dynamics, New Themes*, edited by Ronald Purver, 182–203. New York: St. Martin's Press, 1989.

Rahbek-Clemmensen, Jon. "The Ukrainian crisis moves north. Is Arctic conflict spill-over driven by material interests?" *Polar Record* 53, no. 1 (2017): 1–15.

Rahbek-Clemmensen, Jon, and Gry Thomasen. "How has Arctic coastal state cooperation affected the Arctic Council?" *Marine Policy* 122 (2020): 1–7. https://doi.org/10.1016/j.marpol.2020.104239.

Rahbek-Clemmensen, Jon, and Gry Thomasen. "Learning From the Ilulissat Initiative: State Power, Institutional Legitimacy, and Governance in the Arctic Ocean 2007–1." Centre for Military Studies, University of Copenhagen, 2018. https://cms.polsci.ku.dk/publikationer/learning-from-the-ilulissat-iniative/

download/CMS_Rapport_2018__1_-_Learning_from_the_Ilulissat_initiative.pdf.

Reagan, Ronald. *An American Life*. New York: Simon and Schuster, 1990.

Reagan, Ronald. *The Reagan Diaries,* edited by Douglas Brinkley. New York: Harper Collins, 2007.

Renshon, Jonathan. *Fighting for Status: Hierarchy and Conflict in World Politics*. Princeton: Princeton University Press, 2017.

Richardson, Elliot L. "Jan Mayen in Perspective." *The American Journal of International Law* 82, no. 3 (1988): 443–458.

Robinson, Ronald, John Gallagher, and Alice Denny. *Africa and the Victorians: The official mind of imperialism*. London: Macmillan, 1961.

Rogne, Odd. "Initiation of the International Arctic Science Committee (IASC)." *IASC after 25 Years*. Special Issue of the *IASC Bulletin* (2015): 9–19.

Rosen, Mark E., and Cara B. Thuringer. "Unconstrained Foreign Direct Investment: An Emerging Challenge to Arctic Security." CNA's Occasional Paper series (November 2017). https://www.cna.org/cna_files/pdf/COP-2017-U-015944-1Rev.pdf.

Rossi, Christopher. "The Northern Sea Route and the Seaward Extension of Uti Possidetis (Juris)" *Nordic Journal of International Law* 83, no. 4 (2014): 476–508.

Rossi, Christopher. "'A Unique International Problem': The Svalbard Treaty, Equal Enjoyment, and Terra Nullius: Lessons of Territorial Temptation from History." *Washington University Global Studies Law Review*, xv, no. 1 (2016): 93–136.

Rossi, Christopher R. "Treaty of Tordesillas Syndrome: Sovereignty ad Absurdum and the South China Sea Arbitration." *Cornell International Law Journal*, 50, no. 2 (2017): 231–283.

Rottem, Svein Vigeland. *The Arctic Council: Between Environmental Protection and Geopolitics*. Singapore: Springer, 2020.

Rottem, Svein Vigeland. "A Note on the Arctic Council Agreements." *Ocean Development & International Law* 46, no. 1 (2015): 50–59.

Rowe, Elana Wilson. *Arctic governance. Power in cross-border cooperation*. Manchester: Manchester University Press, 2018.

Sale, Richard, and Eugene Potapov, *The Scramble for the Arctic*: Ownership, Exploitation and Conflict in the far North London: Frances Lincoln, 2010.

Sartre, Jean-Paul. "Existentialism is humanism." In *Existentialism: Basic writings*, edited by Charles Guignon and Derk Pereboom, 290–308. Indianapolis, IN: Hackett Press, 1946 [2001].

Sawhill, Steven G. "Cleaning-up the Arctic's Cold War Legacy: Nuclear Waste and Arctic Military Environmental Cooperation." *Cooperation and Conflict* 35, no. 1 (2000): 5–36.

Sayeed, Fahad. "Sovereign default of Iceland: Voting outcomes of the referenda 2010 and 2011 conducted for the approval of the Icesave bill." *SSRN*, March 13, 2015. https://papers.ssrn.com/sol3/papers.cfm?abstract_id=2571216.

Schmitt, Carl. *The Concept of the Political*. Chicago and London: The University of Chicago Press, 2007 [1932].

Schneider, A. "Northern Sea Route: A Strategic Arctic Project of the Russian Federation." *Problems of Economic Transition* 60, no. 1–3 (2018): 195–202.

Schofield, B. *The Arctic Convoys*. London: Macdonald & Jane's Ltd., 1977.

Scopelliti, Marzia, and Elena Conde Pérez. "Defining security in a changing Arctic: helping to prevent an Arctic security dilemma." *Polar Record* 52, no. 6 (2016): 672–679.

Sellheim, Nikolas, and Dwayne Ryan Menzes, eds. *Non-state Actors in the Arctic Region*. Cham: Springer 2022.

Sergunin, Alexander. *Explaining Russian Foreign Policy Behavior: Theory and Practice*. Stuttgart: ibidem Press, 2016.

Sergunin, Alexander. "Russia and Arctic Security: Inward-Looking Realities." In *Breaking Through: Understanding Sovereignty in the Circumpolar Arctic*, edited by Wilfried Greaves and P. Whitney Lackenbauer, 117–136. Toronto: University of Toronto Press, 2021.

Sharapova, Anna, Sara Seck, Sarah MacLeod, and Olga Koubrak. "Indigenous Right and Interests in a Changing Arctic Ocean: Canadian and Russian Experiences and Challenges." *Arctic Review on Law and Politics*, 13 (2022): 286–311.

Shevtsova, Lilia. *Russia: Lost in Transition. The Yeltsin & Putin Legacies*. Washington, D.C.: Carnegie Endowment for International Peace, 2007.

Shultz, George P. *Turmoil and Triumph: My Years as Secretary of State*. New York: Charles Scribner's Sons, 1993.

Schweller, Randall L. "Bandwagoning for Profit: Bringing the Revisionist State Back In." *International Security* 19, no. 1 (1994): 72–107.

Shvets, Daria, and Kamrul Hossain. "The Future of Arctic Governance: Broken Hopes for Arctic Exceptionalism." *Current Developments in Arctic Law* 10 (2022): 49–63.

Sigurjónsson, Arnór. *Íslenskur her. Breyttur heimur, nýr veruleiki* [An Icelandic Military: New Challenges in a Changed World]. Reykjavík: Arnór Sigurjónsson, 2023.

Sigurjónsson, Jóhann. "Icelandic Perspectives on the Agreement to Prevent Unregulated High Seas Fisheries in the Central Arctic Ocean." *The Yearbook of Polar Law Online* 12, no. 1 (2020): 268–284.

Skidmore, David. "Carter and the Failure of Foreign Policy Reform." *Political Science Quarterly* 108, no. 4 (1993/1994): 699–729.

Smieszek, Malgorzata, Timo Koivurova, and Egill Thor Nielsson. "China and Arctic Science." In *China and Arctic Science*, edited by Timo Koivurova and Sanna Kopra, 42–61. Leiden: Brill, 2020.

Smieszek, Malgorzata. "Evaluating institutional effectiveness: the case of the Arctic Council." *The Polar Journal* 9, no. 1 (2019): 3–26.

Smith, Gaddis. *Morality, Reason, and Power: American Diplomacy in the Carter Years*. New York: Hill and Wang, 1986.

Smith, Julianne, and Jerry Hendrix. *Forgotten Waters: Minding the GIUK Gap*. Washington, D.C.: Center for a New American Security, May 2017. https://www.cnas.org/publications/reports/forgotten-waters.

Snæbjörnsson, Arnór. "Smugudeilan: Veiðar Íslendinga í Barentshafi 1993–1999" [The Loophole Dispute: Icelandic Fishing the Barents Seas, 1993–1999]. MA thesis, University of Iceland, 2015.

Snæbjörnsson, Arnór. "Nýr samningur um að koma í veg fyrir stjórnlausar úthafsveiðar í miðhluta Norður-Íshafsins" [New Agreement on the Prevention of Irregular Fishing on the High Seas in the Central Arctic Ocean], *Úlfljótur*, January 23, 2019. https://ulfljotur.com/2019/01/23/nyr-samningur-um-ad-koma-i-veg-fyrir-stjornlausar-uthafsveidar-i-midhluta-nordur-ishafsins/.

Solski, Jan Jakub. "The Northern Sea Route in the 2020s: Development and Implementation of Relevant Law." *Arctic Review of Law and Politics*, 11 (2000): 383–410.

Staun, Jørgen. "Russia's strategy in the Arctic: cooperation, not confrontation." *Polar Record* 53, no. 3 (2017): 314–332.

Steinberg, E. Philip, and Klaus Dodds. "The Arctic Council after Kiruna." *Polar Record* 51, no. 1 (2015): 108–110.

Stefansson, Vilhjalmur. *Discovery: The Autobiography of Vilhjalmur Stefansson*. New York: McGraw-Hill, 1964.

Steinsson, Sverrir. "The Cod Wars: A re-analysis." *European Security* 25, no. 2 (2016): 256–275.

Steinsson, Sverrir. "Do liberal ties pacify? A study of the Cod Wars." *Cooperation and Conflict* 53, no. 3 (2018): 339–355.

Steinveg, Beate. "Governance by conference? Actors and agendas in Arctic politics." PhD diss. The Arctic University of Norway, 2020.

Steinveg, Beate. "The role of conferences within Arctic governance." *Polar Geography*, 44, no. 1 (2021a): 37–54.

Steinveg, Beate. "Exponential Growth and New Agendas – A Comprehensive Review of the Arctic Conference Sphere." *Arctic Review on Law and Politics* 12 (2021b): 134–160.

Strauss, Lon. "U.S. Marines and NATO's Northern Flank." *Arctic Review on Law and Politics* 13 (2022): 72–93. https://arcticreview.no/index.php/arctic/article/view/3381/6326.

Strong, Robert A. *Working in the World: Jimmy Carter and the Making of American Foreign Policy.* Baton Rouge: Louisiana State University Press, 2000.
Sun, Kai. "China and the Arctic: China's Interests and Participation in the Region." East Asia-Arctic Relations: Boundary, Security and International Relations, Paper 2. GIGI (November 2013). https://www.cigionline.org/publications/east-asia-arctic-relations-boundary-security-and-international-politics/.
Taubman, William. *Gorbachev: His Life and Times.* New York: W.W. Norton, 2017.
Tavares, Rodrigo. "Regional Clustering of Peace and Security." *Global Change, Peace & Security* 21, no. 2 (June 2009): 153–164.
Tessman, Brock F. "System Structure and State Strategy: Adding Hedging to the Menu." *Security Studies* 21, no. 2 (2012): 192–231.
Thorhallson, Baldur and Hjalti Thor Vignisson. "A Controversial step: Membership of the EEA." In *Iceland and European Integration: On the Edge*, edited by Baldur Thorhallson, 38–49. London: Routledge, 2004.
Thorhallsson, Baldur. "Evrópustefna íslenskra stjórnvalda: Stefnumótun, átök og afleiðingar." In *Uppbrot hugmyndakerfis. Endurmótun íslenskrar utanríkisstefnu 1991–2007*, edited by Valur Ingimundarson, 67–136. Reykjavík: Hið íslenska bókmenntafélag, 2008.
Thorhallsson, Baldur. "Iceland: A reluctant European." In *The European Union's non-members: independence under hegemony*? Edited by Erik Oddvar Eriksen and John Erik Fossum, 118–136. London and New York, 2015.
Thorhallsson, Baldur and Pétur Gunnarsson. "Iceland's alignment with the EU-US sanctions on Russia: autonomy versus dependence." *Global Affairs* 3, no. 3 (2017): 307–318.
Thorhallsson, Baldur, and Snæfridur Grimsdottir, *Lilliputian Encounters with Gulliver: Sino–Icelandic Relations from 1995 to 2021.* Reykjavík: International Institute at the University of Iceland, 2021.
Tillman, Henry, Yang Jian, and Egill Thor Nielsson. "The Polar Silk Road: China's New Frontier of International Cooperation." *China Quarterly of International Strategic Studies* 4, no. 3 (2018): 345–362.
Tunsjø, Øystein. *Security and Profit in China's Energy Policy: Hedging against Risk.* New York: Columbia University Press, 2013.
Tunsjø, Øystein. "U.S.-China Relations: From Unipolar Hedging to Bipolar Balancing." In *Strategic Adjustment and the Rise of China: Power and Politics in East Asia*, edited by Robert S. Ross and Øystein Tunsjø, 41–68. Ithaca/London: Cornell University Press, 2017.
Tymchenko, Leonid. "The Northern Sea Route: Russian Management and Jurisdiction over Navigation in Arctic Seas." In *The Law of the Sea and Polar Maritime Delimitation and Jurisdiction*, edited by Alex G. Oude Elferink and Donald R. Rothwell, 277–81. Leiden: Brill, 2001.

Ulfstein, Geir. *The Svalbard Treaty: From Terra Nullius to Norwegian Sovereignty.* Oslo: Scandinavian University Press, 1995.

Van den Berghe, Pierre. "Ethnicity and the Sociological Debate." In *Theories of Ethnic and Race Relations*, edited by John Rex and David Mason, 246–63. Cambridge: Cambridge University Press, 1998.

Van der Linden, Mieke. *The Acquisition of Africa (1870-1914): The Nature of International Law*. Boston: Brill, 2016.

Värynen, Raimo. "Regionalism: Old and New." *International Studies Review* 5, no. 1 (2003): 25–51.

Vylegzhanin, A.N., and V.K. Zilanov. *Spitsbergen: Legal Regime of Adjacent Maritime Areas.* Translated by William E. Butler. Utrecht: Eleven International Publishing, 2007.

Walt, Stephen M. "Alliance Formation and the Balance of World Power." *International Security* 9, no. 4 (1985): 3–43.

Waltz, Kenneth N. *Theory of International Politics.* Reading: Addison-Wesley, 1979.

Wegge, Njord. "Arctic Security Strategies and the North Atlantic States." *Arctic Review on Law and Politics*, 11 (2020): 360–382.

Wither, James Kenneth. "An Arctic security dilemma: assessing and mitigating the risk of unintended armed conflict in the High North." *European Security* 30, no. 4 (2021): 649–666.

Wilson, Page. "An Arctic 'cold rush'? Understanding Greenland's (in)dependence question." *Polar Record* 53, no. 5 (2017): 512–519.

Wilson, Page, and Auður H. Ingólfsdóttir. "Small State, Big Impact? Iceland's First National Security Policy." In *Routledge Handbook of Arctic Security*, edited by Gunhild Hoogensen Gjørv, Marc Lanteigne, and Horatio Sam-Aggrey, 188–197. Abingdon and New York: Routledge, 2020.

Whitehead, Thor. "Austurviðskipti Íslendinga" [Iceland's Trade with the East Bloc] *Frelsið* 3, no. 3 (1982): 198–211.

Whitehead, Thor. "Iceland in the Second World War," Ph.D. diss. Oxford University, 1978.

Whitehead, Thor. *Ófriður í aðsigi* [War Approaching]. Reykjavík: Almenna bókafélagið, 1980.

Whitehead, Thor. *Stríð fyrir ströndum* [War Close to the Shores]. Reykjavík: Almenna bókafélagið, 1985.

Whitehead, Thor. "Leiðin frá hlutleysi [The Road from Neutrality]." *Saga* 29 (1991): 63–121.

Whitehead, Thor. *Milli vonar og ótta* [Between Hope and Fear]. Reykjavík: Vaka-Helgafell, 1995.

Whitehead, Thor. *Bretarnir koma* [The British Arrive]. Reykjavík: Vaka-Helgafell, 1999a.

Whitehead, Thor. *The Ally Who Came in from the Cold: A Survey of Icelandic Foreign Policy 1946–1956*. Reykjavík: University of Iceland Press, 1998.

Whitehead, Thor. *Bretarnir koma* [The British Arrive] (Reykjavík: Vaka-Helgafell, 1999b);

Whitehead, Thor. "Hlutleysi Íslands á hverfanda hveli 1918–1945" [Iceland's Neutrality in Limbo]. *Saga* 44, no. 1 (2006): 21–64.

Wohlforth, William C., Benjamin De Carvalho, Halvard Leira, and Iver Neumann. "Moral authority and status in International Relations: Good states and the social dimension of status seeking." *Review of International Studies* 44, no. 3 (2017): 526–546.

Wolf, Sarah. "Svalbard's Maritime Zones, their Status Under International Law and Current and Future Disputes Scenarios." Stiftung Wissenschaft und Politik, Berlin. Working Paper. FG 2, no. 2 (2013), 1–37.

Woodman, Richard. *Arctic Convoys 1941–1945*. London: John Murray, 1994.

Yanyan, Shan, Jianfeng He, Guo Peiqing, and Liu He. "An assessment of China's participation in polar subregional organizations." *Advances in Polar Science*, 34, no. 1 (2023): 56–65.

Yang, Jian. *Arctic Governance and China's Engagement*. Shanghai: Shanghai Institutes for International Studies, 2022.

Yilmas, Serafettin. "Exploring China's Arctic Strategy: Opportunities and Challenges." *China Quarterly of International Strategic Studies* 3, no. 1 (2017): 57–78.

Young, Oran R. "Whither the Arctic? Conflict or cooperation in the circumpolar north." *Polar Record* 45, no. 1 (2009): 73–82.

Young, Oran R. "The future of the Arctic: cauldron of conflict or zone of peace?" *International Affairs* 87, no. 1 (2011): 185–193.

Young, Oran R. "Can the Arctic Council Survive the Impact of the Ukraine Crisis?" *Georgetown Journal of International Affairs,* December 30, 2022. https://gjia.georgetown.edu/2022/12/30/can-the-arctic-council-survive-the-impact-of-the-ukraine-crisis/.

Østerud, Øvind and Geir Hønneland. "Geopolitics and International Governance in the Arctic." *Arctic Review on Law and Politics* 5, no. 2 (2014): 156–176.

Østhagen, Andreas. "The Good, the Bad, and the Ugly: Three Levels of Arctic Geopolitics." In *The Arctic and World Order*, edited by Kristina Spohr and Daniel S. Hamilton, 357–378. Washington, D.C.: Foreign Policy Institute/ Henry A. Kissinger Center for Global Affairs, Johns Hopkins University SAIS, 2020a.

Østhagen, Andreas. "What is the Point of Norway's new Arctic Policy?" The Arctic Institute, December 2, 2020b. https://www.thearcticinstitute.org/point-norway-new-arctic-policy/.

Østhagen, Andreas. "The Arctic security region: misconceptions and contradictions." *Polar Geography* 44, no. 1 (2021a): 55–74.

Østhagen, Andreas. "Norway's Arctic policy: still high North, low tension?" *The Polar Journal*, 11, no. 1 (2021b): 75–94.

Åtland, Kristian. "Mikhail Gorbachev, the Murmansk Initiative, and the Desecuritization of Interstate Relations in the Arctic." *Cooperation and Conflict* 43, no. 3 (2008): 289–311.

Åtland, Kristian. "Interstate Relations in the Arctic: An Emerging Security Dilemma?" *Comparative Strategy* 33, no. 2 (2014): 145–166.

Zellen, Barry. *Arctic Doom, Arctic Boom: The Geopolitics of Climate Change in the Arctic*. London: Praeger, 2009.

Zellen, Barry Scott. "As War in Ukraine Upends a Quarter Century of Enduring Arctic Cooperation, the World Needs the Whole Arctic Council Now More Than Ever." *Northern Review* 54 (2023): 137–160.

Zellen, Barry Scott, ed. *The Fast-Changing Arctic: Rethinking Arctic Security for a Warmer World*. Calgary: University of Calgary Press, 2013.

Zhuravel, Valery P. "Arctic Council: Outcome of the First Year of the Icelandic Presidency."

Þórhallsdóttir, Sigurveig. "Norður-Íshafssamningurinn" [The Central Arctic Ocean Agreement]. MA thesis, University of Iceland, 2020.

Index[1]

[1] Note: Page numbers followed by 'n' refer to notes.

© The Author(s), under exclusive license to Springer Nature Switzerland AG 2024
V. Ingimundarson, *Iceland's Arctic Policies and Shifting Geopolitics*, https://doi.org/10.1007/978-3-031-40761-1

H

I

Printed in the USA
CPSIA information can be obtained
at www.ICGtesting.com
CBHW062246291124
18248CB00004B/23